RAPPELLING

by *TOM MARTIN*

RAPPELLING

First Edition — May, 1985
Second Printing — March, 1986
Edition II — March, 1987
Edition II, Revised — February, 1988
Second Printing — December, 1988

Library of Congress Card Number 87-92014

ISBN 0-930871-02-2

Published by SEARCH
106 Sterling Ave., Mt. Sterling, KY 40353

DEDICATION

This book is dedicated to the countless people in rescue squads, fire departments, police departments, and the military who rappel to save lives. May it make your job safer.

During the preparation of this second edition, many people generously shared their knowledge, time, and equipment. Listed below, with greatest gratitude, are those who helped the most.

Bill Allfather
Laura Berkley
Russ Born
Will Brown
Veronica Burgess
Larry Caldwell
Martha Carlson
Thomas Crucet
Steve Davis
Joe Dillon
Don Fig
Bill Forrest
Martin Hackworth
Wanda Helm
Tom Hunt
Norman Lawson
Becky Long
Larry Martin
Juergen Meschke
Jack Sawicki
Geary Schindel
Bruce Smith
Jerry Smith
Gary Stidham
Gary Storrick
Tim Thorne
John Weinel

A special thanks to my Lord, Jesus Christ, for preserving us thru countless rappels.

TABLE OF CONTENTS

FOREWORD

On the television program *All In The Family*, there was a scene in which a mentally handicapped guy showed Archie Bunker a plaque inscribed with these words: "Every man is my superior, in that I may learn from him." If you do not understand these words, this book will prove of little value to you.

This book is the result of twenty years of research, study, and rappelling. Like the first edition, it is being published in the hope that it might, in some small way, help curb the number of injuries and deaths which occur during rappelling. Rappelling can, and should, be a very safe activity. Most deaths are the result of stupid mistakes or inferior equipment. Both problems can be eliminated if the rappeller takes the time to study the mistakes of others, use only quality equipment, learn to safely use this equipment, and **THINK** as he rappels.

The intent of this book is not to encourage rappelling – or discourage it. People are going to rappel, whether from desire or necessity. It is simply hoped that the information contained herein will make this activity both safer and more enjoyable. The methods and equipment recommended should not be construed as "the only way." You, the reader, should use this information as a foundation on which to build, along with your own experiences and logic, a system of rappelling which best suits you and your situation.

During your ongoing learning experience, don't let any "experts" convince you that their way is the only way, and don't let anyone talk you into doing anything which seems stupid or illogical. Think for yourself! When going rappelling, leave most of your ego at home. There is nothing "macho" about being able to rappel. Children and chimpanzees can be taught to rappel!

Although great amounts of time and effort were spent in the production of this book, remember that it is no substitute for a competent instructor. This book cannot tell you if a knot has been tied wrong, or if an anchor may fail, or if you are using the best method in a given situation. An experienced instructor can. Just never forget that even the most competent instructor can be wrong. Before you commit yourself to a rappel, check everything yourself – your life is at stake.

As you gain knowledge and experience, you will find that there is probably nothing about rappelling which can't be improved upon. By making the improvement, you might be the means of saving a life or preventing an injury. As you let your creative energies flow, just **THINK, STUDY, TEST, and BE CAREFUL**. And if you come up with something of value, share it – with your instructor, with those you rappel with, and with the author of this book. If we all pool our knowledge, rappelling will be made safer, and lives will be spared. (The author can be contacted at the SEARCH address, in the front of this book.)

INTRODUCTION

Rappelling, or Abseiling, is the science of sliding down a rope. It is a skill used by mountain climbers, firemen, soldiers, cavers, police, thrillseekers, and countless others. It is a way to descend a rope in a safe, controlled manner, in which one has to support only a small fraction of his total weight with his hands. Over the many years that people have been rappelling, this science has evolved from crude body rappels to a highly sophisticated and safe technology.

The history of rappelling is somewhat obscure, probably reaching into the distant past when some seaman found that descending a rope could be made easier by winding it around a leg or arm. Mountaineers, in the 1800's, developed more sophisticated methods of body rappelling; then, in the middle part of this century, climbers and cavers began developing mechanical descending devices. The use of mechanical descenders is now almost universal amongst the rappelling fraternity.

Rappels have been done from cliffs, buildings, helicopters, towers, and about everything else – and from all heights. The present world rappel record is 3,280 ft. (1 km.) down the side of Mt. Thor, in northern Canada.

Rappelling is a fantastic experience! The thrill of sliding down a rope, in full control of your speed, is much like that of skydiving, and it takes many rappels before the thrill wears off. Part of the initial thrill is fear – at least for most people. After enough rappels, this usually wears off too. Just how safe or dangerous rappelling is, depends on how you wish to make it. It can – and should – be far safer than driving to the area where you rappel. How safe it is depends almost totally on you – the rappeller. You are allowed no mistakes!

For those who've never rappelled before, the rest of this chapter is a brief, step-by-step scenario of what its like. This story will hopefully convey the feelings experienced during an actual rappel, and help you decide if you wish to try. As you read it and find unfamiliar words, consult the glossary and index for a better understanding of what's going on. As you read, remember that this is a positive scenario, with a happy ending. Because rappellers don't always use good judgment, many real rappels don't end so pleasantly.

"Its a beautiful day! Lets go rappelling. Our friends have arrived, and its time for a gear check. Each rope is carefully inspected, by hand, down its entire length. Once that is done, the rest of the gear is given a similar inspection. We also check to make sure there are enough carabiners and accessory cord – and that everyone has a helmet and gloves. Everything is now loaded on board, and we head out.

A few miles down the road we come to a good cliff, with few trees and obstructions near the base. One side of it is only about eight feet high; a good place to start beginners. The other side is over two hundred feet high; our "experts" will have fun there. We unload the gear and head for the low side, so that we can train the new people.

A good place to rappel is soon found. It has a couple of sturdy trees a few feet from the edge, and the edge is firm and nicely rounded. Because some of our group have no rappel experience, it seems wise to hook up a safety belay rope – in case someone should have problems during the rappel. The belay will make us "old-timers" feel better too. After all, the unexpected can happen. Better safe than sorry!

The group leader inspects the edge and a nearby tree and determines them both to be safe. The rappel rope is now uncoiled and looped around the tree. Protectors are put on the rope, and it is let down. The leader checks to see if both ends reach the ground. By looping the rope around the tree, we have two lines to rappel on. Rappelling on two lines, instead of one, almost doubles our margin of safety.

Now that the rappel lines are set up and checked out for safety, its time to rig the belay system. A strong sling is placed around the tree and connected to a belay device. A rope is then run thru this device; to be attached later to the rappeller's harness. This is called a top belay, and will hopefully keep a rappeller from falling, in the event his rope should get cut or he should lose control during the rappel.

The leader also sends an experienced person to the bottom of the cliff. He will provide a bottom belay and help guide the new rappellers down. He wears his safety helmet, in case any loose sticks or boulders should come down.

Whoever owns the rope and rigs the anchor gets to go over first! This tradition helps maintain equipment quality.

The first rappel of the day goes smoothly. The rope and belay system seem safe. Its time for the new rappellers to try. Its now your turn!

With the aid of the leader, you slip into the rappel seat harness. Both of you – together – check to make sure all knots, buckles, and etc. on the harness are properly fastened. Also, check to make sure the harness webbing is not frayed or cut, and that all stitching is secure. Next, the descender is connected to the harness with a sturdy locking carabiner. Make sure you lock the carabiner's gate.

The belay rope is now connected to your harness, and to a safety loop tied around your waist. This loop, made of wide webbing, is simply extra protection in case the harness should fail. The belayer then takes most of the slack out of the belay rope.

Next, put on your helmet; making sure the chin strap is secure. Put on your gloves, and with the leader's assistance, connect the descender to the rope. The braking end of the rope, below the descender, should be held in which ever hand you are most comfortable using. Make sure the descender is connected so as to provide enough friction. If you have too little, you will have a hard time holding your weight with your braking hand. If you have too much, you may not be able to slide down the rope. Since the rope has been looped around the tree for a double-line rappel, be sure the descender is connected onto both lines!

Its now time to go over. Once again all equipment and systems are given a thorough inspection. If all is well, the leader OK's the rappel, and the belayer signals that he is "On Belay." Since you will be rappelling down "tail first," start carefully backing toward the edge; sliding both rope protectors down the ropes, above the descender, with your balance hand.

Before you get to the edge, stop yourself with your braking

hand and put full pressure on the rappel system. This will help you determine if the descender's friction is correct – and if the anchor will hold.

When you get to the edge, look it over well. Your life is at stake here! Don't trust the leader's instruction without first checking things for yourself. Make sure the edge is not sharp, lest your rope get cut as you go over. Don't rely totally on protectors to protect your rope. Make sure both ends of the rope reach the ground.

Looking down, even tho the drop is not high, you feel a bit scared. Perhaps your legs are shaking. Don't worry, this is normal for beginners; the fear will go away in time. You are using the best equipment, and have competent instructors. If you do things correctly, there should be little danger.

But don't get overconfident yet! Accidents can happen. Be on your guard at all moments. Think every move thru carefully. Also, now is a good time to take stock of your life, and make sure it is in order. You are involved in a very safe activity, but there is always the danger that you may die!

You've thought it thru, and got the fear under control, and you are now ready to rappel. Signal the belayer that you are "On Rappel." He will signal, once again, that he is "On Belay." Place your feet squarely on the edge, with your legs slightly spread, and begin leaning back. If you are having a hard time maintaining balance, spread your legs even more.

As you lean back, keep your knees bent slightly for balance. Keep pulling the rope protectors down the rope as you go, and place them so that they will lay on the edge of the rock when the rope makes contact with it. Lean back until you are "standing" horizontally on the vertical cliff wall. You are now hanging from the rope, and can begin walking backwards down the cliff. The rope is holding, so don't be scared. Your weight, as you walk down the cliff, should be on the soles of your feet – not on your toes or heels. Walk down the wall as you would back across your floor at home.

Note that you can control your speed by how tightly you grip the rope with your braking hand. Anytime you wish to stop, you can, just by gripping harder. As you slowly slide down the rope, you are in full control. The belayers are there only for your safety. Take time to look around at the scenery and experience the joy of the rappel.

Note that once you got over the edge, most of your fear disappeared. The rope is holding your weight quite nicely. Rappelling is actually fun!

Now that you are really rappelling, slide a little bit faster, then slower, so that you can get the feel of things. When you go off higher cliffs, later today, you will have more distance to rappel, and may need to go a bit faster than you are going now. As you near the ground, try lightly springing and bouncing off the cliff wall. This will help you get the feel of bouncing over holes and rough spots on bigger cliffs.

You are almost ready to crash on top of your bottom belayer, so it is best to slow the rappel. He would not like being squished! After he releases the rope, slide to the ground, remove your descender from the rope, signal to the top belayer that you are "Off Rappel," and give thanks for a safe journey. You've done a fine job!"

If you've never rappelled before, you may still be confused about how to rappel – even after reading this detailed scenario. As you work with your instructor, and gain knowledge and experience, these things will become second nature to you, and in time you may be telling it to others – as their instructor. For now, don't worry about it. Study the rest of this book, every word of it, and prepare yourself for your first – safe – rappel.

BASICS

Basically, rappelling is a safe way to come down a rope. Knowing the basics of rappelling will get you by, but knowing the basics – and nothing more – can also get you killed. In the art of rappelling, a little knowledge can truly be a dangerous thing.

There is much physics and mathematical data dealing with strength and heat dissipation capabilities of rappel equipment. There is a place for such data, but not in this book. The object here is to convey basic, factual, practical data which can be readily understood and utilized by all rappellers.

The most basic of all methods of rappelling are body rappels, as described in Chapter 4. They are not a pleasant way to go, but every rappel student should learn them first. They teach discipline and self-control, and make one grateful for safer and less painful mechanical methods. They could also save your life in a survival situation!

After body rappelling is mastered, start learning about the mechanical methods of descending. Learn as much as you can about the virtues and faults of each system, and experiment with as many varieties as you can. Eventually you will determine which systems and devices are best for your needs. Above all, learn to tie knots – correctly! You won't live long if you don't.

As stressed in the Foreword, only the finest rappel equipment should be used. If you are too poor – or too cheap – to go first class, then don't go at all. The consequences could be fatal! Quality equipment is usually expensive, but price doesn't always reflect quality. Some of the most expensive is true garbage, as is some of the cheaper stuff. Buy only equipment which has a proven history of reliability, and as much possible, buy equipment which meets the safety standards set up by the various climbing and rescue organizations and government agencies. Always know what you are buying – before you buy.

Knowledge of every aspect of rappelling is necessary to prevent accidents, but knowledge alone won't cut it. Commitment to safety is what separates the living from the dead – or permanently injured! In the interest of safety, the rest of this chapter is a long, boring list of do's and don'ts. They aren't all absolute rules, but it might be wise to take them that way.

Start low. This advice is good for beginners and instructors. Start with a drop no higher than your head, and learn the basics there. Better yet, start by rappelling down a gentle incline.

Be Cool. Fear can kill – or it can save, it depends on the rappeller. Some beginners have no fear at all; they will go over an edge with little thought or preparation. These are dangerous people! They tend to kill themselves and others. Moderate fear is good for a beginner. It helps prevent mistakes, and improves concentration. The only fear which is dangerous is that which overwhelms the mind, and causes panic. That kind of fear can kill! So long as the mind can function logically, fear is not dangerous. You can be so scared that your legs are shaking, but you can still do a safe rappel – if you have the determination. Determination and practice will eventually banish the fear. Be cool, but not too cool!

Keep off the chemicals. If you're the kind of folk who drives or climbs while under the influence of consciousness-altering chemicals, then don't waste your time studying this book. Give your copy to the person who will rescue your tail when you fall!

Always check equipment. Equipment does age, and it does break. Becoming aware of a defect as you are falling toward the ground is a stupid way to go. This check must include what you tie your rope onto – your anchor. Also, make sure there are no sharp burrs or scratches on the descender, which might damage the rope.

Never trust strength ratings – until you know how they were determined. Some manufacturers rate their equipment by its minimum breaking strength; some rate it by maximum strength; others simply take an average. Minimum strengths are the most honest, and are the only ones to trust your life to. Before you believe a strength rating, learn where it came from.

Tie yourself in. When you are setting up the rappel, it is sometimes wise to tie yourself to a rigid anchor; particularly if the area is very sloped or wet or icy. As the Bible puts it: "Let him that thinkest he standeth take heed lest he fall." KJV, I Corinthians 10.12.

Never step on a rope. The rope is your lifeline – if it breaks, your life may be over. Stepping on a rope forces grit into the fibers. The rope can also be cut – if there should be sharp rocks or broken glass beneath it.

Rappel double-line. The Army always rappels double-line (on two ropes). They have perhaps the best safety record of any group teaching rappelling. Its not always practical to use two ropes, particularly on long drops; but this should be the rule when possible. Double-line rappels save lives!

Do it reach the ground? Making sure the rope reaches the ground may seem like stupid advice; but its amazing how many people wind up rappelling off the end of a rope – far, far from the ground. The problem is especially bad during night rappels. If you can't see the end of your rope, pull it back up and tie the end onto your harness carabiner – or better yet, go home. This is the main reason to carry ascending equipment.

Use a protector. Until you've seen a rope cut before your very eyes, you cannot realize just how fragile they are. Even a dull rock edge can cut a rope under the right circumstances. Always protect the rope with a protector or padding. The rope will last longer and so will you!

Wear your gloves and helmet. A quality helmet and gloves are of extreme importance. If something goes wrong during a rappel, gloves can literally mean the difference between life and death. If you've ever had an 8″ boulder whiz by your head while rappelling, then you already know the virtue of helmets.

Watch your feet! The rock you accidentally dislodge on top can accidentally kill someone below!

Run the rope thru the harness carabiner. By running the braking end of the rope thru the harness carabiner, you may be able to maintain some control if your descender should break, or come off the rope. This isn't practical or necessary with some descenders, but it is with others. Just make sure the rope doesn't rub across the harness and cut it. There are better ways of securing a rappel, and they will be described later.

Have enough friction. Use enough friction in your rappel system so that you don't have to rappel with a "death grip" on the rope. In normal rappels, it is advisable to have enough friction so that you can safely descend using only your thumb and trigger finger.

Keep your braking hand on the rope. With most descenders, if you let go of the rope you will start sliding to the ground. If something happens during the rappel, don't remove your braking hand from the rope – train yourself to grip harder. Letting go with the braking hand, and grabbing onto the rope above the descender, is a good way to turn a small accident into a big one.

Fire, Fire! If you happen to have a carbide light on your helmet, be careful. Rope burns!

Belays are good for your health. All beginners must be belayed. Everyone should, but it isn't always practical or possible. Just bear in mind that good belays have saved lots of lives.

Don't show off. Fast rappels and spectacular bounces can be done by experts; but probably shouldn't. Beginners must always rappel slowly and smoothly. Bouncing off walls as you go down is very thrilling. So is falling – when the bounces cause your rope to be cut! Fast rappels are fun too; if your rope doesn't melt apart, and if you don't smash into the ground because of unexpected rope stretch during braking.

Never loan a rope to anyone – is a good rule. For sure, only loan your rope to experienced rappellers who can be trusted to take care of it. You owe it to yourself, and those you rappel with, to maintain safe equipment and fully know its history of use. Loaning ropes to neophytes is a good way to reduce your life span!

Keep out of waterfalls. Rappelling down a waterfall can be great fun – unless the flow is so great that you can't swing out. Don't count on rappelling quickly to the bottom. When ropes get wet, their friction coefficients rise. Also, things do jam. If you don't like drowning, stay out of the water!

Be careful ladies! Breasts sometimes get caught in descenders. Always carry prusiks for getting unjammed.

Beware using ropes of different diameters. Tying a big and a small rope together, and rappelling double-line, can be hazardous; particularly if they are looped thru an anchor sling. The small rope stretches more than the large one, and can saw thru the sling as you rappel. It can also break easier. To make things safer, put a descending ring on the sling – and rappel gently.

Use big ropes. Large diameter ropes are stronger than small diameter ropes; they also stretch less and last longer. Just don't go so big that the size and weight become cumbersome. This could lead to accidents.

Don't rappel alone. The reasons are obvious!

Throw it away. Carabiners and descenders, which have fallen long distances onto hard surfaces, may be unsafe. Stress cracks can be created by such impacts, and the metal can be greatly weakened. Better to buy new equipment than risk your life.

Keep good records. Knowing how your equipment was used, and for how long, can spell the difference between safe rappelling and dangerous rappelling. If the equipment is owned by a group, record keeping is especially important.

Look out for holes. As one rappeller was bounding down a cliff, his leg got jammed in a hole. Snap, crackle, pop!

Back it up. Two ropes are safer than one. Two carabiners are safer than one. Two anchor slings are safer than one. Two descending rings are safer than one. Two belays are safer than one. As much as practical, back everything up.

Leave word. Always let someone know where you are going rappelling, and when you'll be back.

Use care when top belaying. Top belays are the safest belays, but they do have problems. They can dislodge rocks, and the rocks can injure those below. Falling rocks have also been known to cut rappel ropes.

Don't rappel during storms. Wet rope conducts lightning.

Tie low. If you have to use a small tree for an anchor, tie on to it as near to the ground as possible. Better to crawl over the edge than fall over the edge!

Keep gear out of little hands. Don't let children play with rappel equipment – particularly things like ropes and harnesses. They can damage themselves, or they can damage the gear.

Destroy the remains. When you retire a piece of equipment, destroy it, or at least mark it so that it won't get mixed with the reliable gear. Rappelling is dangerous enough – without rappelling on junk.

Do it right. A belayer is useless if he loses control. Belayers should normally be tied to a safe anchor.

Don't get hung. It's best not to have clothing or slings around your neck. If such things get pulled into the descender, the rappel can turn into a hanging.

Check knots. Some knots come undone during repeated loads on the rope. Check the knot – before you rappel.

Think your way out. If you drop your harness over the edge, and there is no way out but down, remember that you can use the end of the rappel rope to make a harness. A piece of the rope can also be used for an anchor sling. Just make sure there's enough rope left to reach the ground.

Things happen! Washing machines have been known to damage ropes. Strong anchors sometimes break. Sticht plates have actually cut ropes. Be aware, learn all you can, and be careful.

Pull Carefully. During retrieval of double-line ropes, make sure the lines are not twisted – before you begin pulling. Also be sure there are no knots or kinks in the lines. When pulling, pull slowly and smoothly.

Seek instruction. If you are a beginner, get some training from a competent rappel instructor. Learning to rappel on your own can be deadly.

Don't rappel during hunting season. Someone might mistake you for a turkey!

ROPE

INTRODUCTION

Rope is the heart of a rappel system – rappelling is impossible without it. Except for the rappeller, the rope is the most important part of the system, and must be regarded accordingly. It has also been an indispensable part of the development of man. For at least 5,000 years we have been twisting fibers together to make rope. During an archeological excavation of Indian cliff dwellings in Kentucky, some short pieces of ½ inch, 3-strand laid rope were found. Though brittle with age, they looked as good as hemp rope from a modern hardware. Radiocarbon dating indicated them to be almost 2,000 years old.

Almost any material which is flexible and strong can be used as rope. Although most rappel ropes are made from nylon, special situations may require using some other materials. Because your life is suspended on the rope, it is important to know its strengths and limitations. It is also imperative that you know how to care for the rope; for it is how it is treated that mainly determines its life and safety limits.

CONSTRUCTION

Early ropes were constructed from natural fibers, which are usually fairly short. To have a rope of any length, it was necessary to twist great numbers of fibers together end-to-end. This is done by simply laying the fibers together and twisting them into a tight bundle called a yarn. When such line is put under load, the twisted fibers are crushed together and do not slip apart easily. The fibers are placed at different positions along the length of the yarn, so that when it breaks, it is as apt to break at one place as another.

To keep the fibers from untwisting, two or more yarns are twisted together, generally in a clockwise direction, into a strand. If a stronger rope is needed, several strands are twisted together in a counterclockwise direction. This is standard rope manufacturing procedure, and is called "right-laid" (Fig. 3-1). For making "left-laid" rope, the process is reversed.

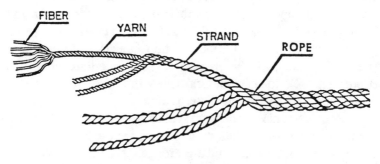

Fig. 3-1

The angle of the spiraling in the yarn and strands determines many characteristics of laid rope; such as strength, elongation, spliceability, stiffness, and etc. The tighter the helix, that is, the greater the number of spirals (or "picks") per inch, the weaker the rope and the greater its stretch or elasticity. In other words, straight fibers offer maximum strength and minimum stretch. However, in order for most ropes to be elastic enough to be safe, fibers must be spiraled.

Very tight spiraling is called "hard lay" and causes the rope to be very stiff, yet very elastic. "Soft lay" ropes have opposite characteristics. "Standard lay" ropes are somewhere in between. Most laid rappel ropes are of the "hard lay" type, so that they can safely absorb sudden loads generated during the rappel. The tightly twisted bundles also resist the effects of dirt and abrasion better than "soft lay."

Laid rope has been constructed of natural fibers like grass, manila, sisal, hemp, cotton, jute, flax, silk, and etc. Manila is made from the leaf stem of the Abaca plant, and is the natural material most commonly used in rope manufacturing. Sisal is next, and then hemp. Silk would be the most desirable material if it were not for cost and availability. Some early climbing ropes were actually made of silk.

Although natural ropes have always been used as rappel lines, they are not desirable; being safe only if new and dry, and then only marginally. Manila loses 1% strength per year of storage, and if not stored under perfect conditions will quickly rot. When wet, natural ropes can absorb their own weight in water and become very heavy; and if cold can freeze. They are destroyed quickly by descenders, and have little resiliency. Manila ropes stretch only about 13% before breaking, and individual fibers will stretch only 2% to 3%. Natural ropes should be used only in emergencies, when nothing better is at hand.

Laid ropes are also made from synthetic fibers. Synthetic fibers are many times longer than natural ones, are usually continuous for the length of the rope, and are stronger and superior in almost every way. Most laid rappel rope is made from Type 6 or Type 6-6 nylon, and is very reliable. The rappel ropes used by the U.S. Army are of laid nylon.

Although laid rope can be used for satisfactory rappels, there are three negative factors in its use. One is the fact that it is of twisted construction; therefore tends to untwist as you rappel. This causes kinks to develop below the descender, which can be troublesome to undo – particularly when rappelling double-line. The problem of the ropes twisting together can be solved by having someone at the bottom keep the lines apart. Kinking can be minimized by using short lines, rappelling slowly, and using descenders which allow the rope to run straight.

The most dangerous property of laid ropes is their great amount of stretch, or elongation. Most will stretch 8% to 12% from the weight of the average rappeller. This can cause accidents when going over certain edges (see Chapter 11). Great amounts of stretch can leave you without a rope. There is a tale about a fellow who rappelled down a deep pit, and went about to explore. When he returned to the rope to climb out, he found the end high above him. The rope had returned to its original length. Stretch can also be lethal when trying to quickly stop near the end of a long, fast rappel. It can allow you to smash into the ground even tho you are not moving down the rope.

One other bad thing about laid ropes is the noise they make when used with most descenders. The loud clicking sounds produced during the rappel could result in the undoing of certain tactical missions. Ropes with smooth exteriors allow much quieter rappels.

On the positive side, laid ropes are strong and reliable, and their dynamic, rubber band effects, make them less apt to pull out anchors or break under sudden heavy loads. A quality synthetic rope will stretch over twice its length before breaking.

Instead of being twisted, yarns can be braided (Fig. 3-2A) or plaited (Fig. 3-2B). Eight-strand plaited rope is perhaps superior to braided rope for rappel purposes, but both types exhibit less stretch than laid rope. They also have greater strength for a given diameter, and less tendency to twist and kink. However, because these ropes are softer than "hard lay," they tend to lose strength faster during use. As in laid rope, every fiber in a braided or plaited rope will come near the surface if the rope is quite long. As the surface of the rope gets abraded, every fiber may be cut at some point along its length. Because of this, and other factors, braided or plaited rappel ropes should be retired when they show the slightest evidence of wear.

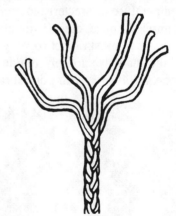

| Fig. 3-2A | Fig. 3-2B |

The finest construction available for rappel rope is called kernmantel. It consists of a core of fibers covered by a braided fiber sheath (Fig. 3-3). The prime purpose of this sheath (mantel) is to protect the fibers of the core (kern) from dirt and abrasion, thus increasing the life of the rope.

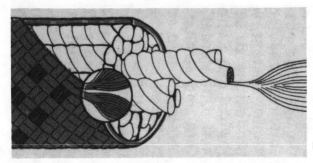

Fig. 3-3

Edelrid

The kern of a rope can have fibers which run straight and parallel, or they can be twisted or braided. As can be seen in Fig. 3-4, there are many ways to make kernmantel. Each method has characteristics which the manufacturer deems important.

Mantel and kern construction, and their mutual relationship, determines many rope characteristics; such as elasticity, gripping quality, knot strength, flexibility, edge strength, high and low temperature characteristics, repeated load capability, life span, mantel slippage, etc. If the mantel is tightly braided about the kern, as in most rappel ropes, great resistance to abrasion results. If the mantel is loosely braided, as in the case of double-braid rope, a soft rope with poor abrasion resistance is produced.

Fig. 3-4

If the mantel is made of different material than the kern, matters are affected even more. A polyester mantel over a nylon kern yields a rope with great resistance to sunlight. A nylon mantel over a polypropylene kern provides strength, plus the ability to float. However, because of the different elongation characteristics of the materials, such ropes may exhibit internal abrasion problems under certain circumstances. If synthetic fibers are combined with a steel wire kern, very unique properties result. Such a rope will not cut easily, but has to be rappelled on very carefully because of the inelasticity of the steel kern. Although a steel kern will not melt easily, the nylon of the mantel will, and can cause descenders to jam during a rappel.

Kernmantel rappel rope can be divided into two classes: Static and Dynamic. Static ropes have kerns of nearly parallel fibers which are held together by a tightly braided mantel. These ropes stretch less than 2% at loads of 200 lbs. (90.7 kg.), and less than 20% at breaking loads. Because of this low stretch, static ropes are very good for safely getting over edges, for long rappels, and for lowering or raising heavy loads. Kinking is also minimized.

Dynamic ropes have kerns in which the fibers are braided or twisted. Their main use in mountaineering and rescue is as belay lines, so they are designed to absorb the high energy dynamic forces generated by a falling climber, thus preventing injury. When put under great stress, they act like rubber bands and perform like laid ropes, with breaking elongations of over 40%. However, at bodyweight (176 lbs. or 80 kg.) they stretch only about 4% to 8%. Kerns consist of one or more bundles of fibers, which may be either braided or twisted. These variations, along with things like fiber diameter and cross sectional shape, are what determines a rope's strength/stretch characteristics.

Dynamic climbing ropes make perfectly good rappel ropes – if the height of the rappel is not great. Their main virtue is the ability to absorb the dynamic loads generated by jerky rappels, which might otherwise pull out or break an anchor. Their main danger is stretch when going over an edge (see Chapter 11). This stretch also makes them undesirable for raising or lowering heavy loads during rescue situations.

If a rope is to be used only for rappelling, particularly for rappels of several hundred feet or more, static rope is the most practical choice. The low stretch makes for safer and far more trouble-free rappels. If a rope is to be used for both rappelling and climbing, dynamic type is usually the way to go. Because of its low stretch, static rope can cause serious injury if used to stop a falling climber – even when using dynamic belay devices and long rope lengths. Because of the high impact forces generated during a fall, belaying with a static rope might also cause anchor failure. A good rule for choosing belay ropes is this: If the anchor point is above you, static rope is generally safe – if the anchor is below you, use dynamic rope.

The prime problem with kernmantel rope is the mantel slipping down the kern. This causes some strength loss and handling problems. However, with most modern ropes this slippage is minimal. Certain descenders cause mantel slippage more than others.

When constructing ropes, it is possible to have each fiber continuous (without splices) for the length of the rope. The best quality ropes are constructed in this fashion. However, many ropes are made from fibers which contain splices, either in the form of knots or long overlaps. This permits construction of very long lengths at low cost. Some rope manufacturers even splice yarns. Splicing of fibers and yarns, if held to a minimum, does not reduce the strength of a rope – few ropes ever break at or near a splice. However, after much use, ropes with overlap type splices can become misshapen because of slippage of the splices. Ropes with no splices have less potential for such problems, and do project a better image of quality.

The construction of quality ropes, which are free of defects, is an extremely complex and demanding job. Most ropes designed for rappelling are made with precision and care – and with the knowledge that any defect will probably cost a life. Ordinary ropes are not constructed with this in mind. **NEVER RAPPEL ON ANY ROPE NOT DESIGNED FOR RAPPELLING!**

MATERIALS

The following is a list of synthetic materials presently used in rope manufacturing. The specifications are general and are presented only to help rappellers understand the limits of materials used to construct ropes. To prevent the reader from overestimating rope capacities, most specifications, such as temperature ratings, are minimums. Of the materials in the list, only nylon is recommended for general rappelling. Ropes made of other materials must be used with caution – in some cases, extreme caution – and should be used only in special rescue or emergency situations. For exact data on rope materials, consult the manufacturer of your rope.

NYLON – In the 1930's Nylon 66 became the first synthetic material to be widely used in ropemaking. It is the material of choice for rappel ropes because it has better abrasion resistance than any other practical rope material. Most synthetic ropes in use today are made of either Type 66 or Type 6 nylon. Type 6, known in Europe as Perlon or Enkalon, is slightly stronger than 66, but has a lower melting point. Type 6 melts at 383° F. (195° C.), whereas Type 66 melts at 446° F. (230° C.). Strength loss can occur at continuous temperatures of 175° F. (79° C.) for Type 6, and 270° F. (132° C.) for Type 66; although one manufacturer of high quality climbing ropes says that his ropes will experience slight strength degradation after several minutes of exposure to temperatures as low as 140° F. (60° C.). At extremely low temperatures, nylon ropes tend to become stronger and less elastic, and have been shown to withstand -94° F. (-70° C.) without damage. However, great care must be exercised when using ropes at such low temperatures.

Under the most ideal conditions of storage, nylon ages very slowly, but it loses strength quite rapidly when exposed to strong ultraviolet light or atomic radiation. It is totally resistant to rot and mildew, but is degraded rather quickly by contact with rusting iron or steel. It is quite resistant to most common chemicals, but is weakened or destroyed by various concentrated acids and bleaching agents, particularly at elevated temperatures. Certain oils, such as lard, can also lower the strength of nylon. When saturated with water, nylon rope can absorb up to 9% into its fibers. The water splits up the hydrogen bonds between groups of nylon molecules, causing about 15% strength loss. Water also causes nylon ropes to shrink and become more stiff.

Nylon can be dyed any color, and some colors increase fiber life because of ultraviolet blocking. Dying fibers after they are manufactured has little effect on strength. However, strength is sacrificed if the dye is mixed into the nylon during manufacture. Also, the strength of a given type of nylon varies greatly depending upon the manufacturer and process used.

POLYESTER – Also known by brand names such as Dacron or Fortrel, this material is used for making very low stretch static rappel ropes. Individual fibers of polyester stretch 11% to 13% at breaking, compared to 16% to 20% for nylon. The breaking strength of polyester varies with its composition and manufacturer, with high quality polyester and nylon ropes having approximately the same strength. Polyester does not absorb shock loads as well as nylon, so has to be rappelled on more carefully. Polyester melts at 470° F. (243° C.), but has been shown to withstand 212° F. (100° C.) for 20 days with no observable strength loss. It can withstand the same low temperature as nylon.

Polyester exhibits far better chemical resistance than nylon, except for strong alkalis and ammonia, and can actually become stronger upon exposure to certain chemicals. It also has far greater resistance to ultraviolet and atomic radiation than nylon. When saturated with water, polyester absorbs only about 1%, and will not rot or mildew.

POLYETHYLENE – This material is widely used to make ropes which float on water. It has approximately half the strength and energy absorption capacity of nylon, so has to be used with extreme caution for rappelling. This is doubly so because of its low melting point (248° F. (120° C.). Polyethylene will not rot or mildew, and does not absorb water.

Pure polyethylene is degraded very quickly by ultraviolet, but this can be remedied to some extent by incorporating various ultraviolet inhibitors during manufacture. Unlike most other materials, polyethylene is generally unaffected by exposure to atomic radiation. This makes it potentially valuable in rescue operations necessitated by nuclear accidents. Except for a few concentrated acids, polyethylene is far more resistant to chemical degradation than nylon or polyester.

POLYPROPYLENE – This material is chemically similar to polyethylene. It is extremely resistant to chemical attack, and is actually less water-soluble than glass! Like polyethylene, it floats, possesses half the strength of nylon, and is degraded quickly by ultraviolet unless protected by inhibitors. Polypropylene melts at 311° F. (155° C.). Because it has better elastic properties than polyethylene, it can be considered a safer material to rappel on.

NOMEX – This unique material is used to make ropes which are fire resistant. It does not melt, but simply decomposes at temperatures over 700° F. (371° C.). At room temperatures it has half the strength of nylon, similar chemical properties, and greater elastic elongation under stress. It should be used with caution for rappelling.

FIBERGLASS – Ropes are occasionally made of glass when maximum fire and chemical resistance is necessary. It is useful to temperatures of nearly 940° F. (504° C.), and is resistant to degradation by light, radiation, and all chemicals except for hydrofluoric acid. Fiberglass ropes have little elasticity, and any sharp bends or knots can bring on their catastrophic failure. They should never be used for rappelling except in emergencies, and when used should be rappelled on slowly, with no jerks or fast stops. It is also imperative that the rope not be bent at sharp angles while passing thru the descender. Spools are excellent descenders for use with brittle ropes.

KEVLAR – This synthetic is one of the strongest materials in existence, being stronger than steel and stiffer than glass. It is used to make everything from "bullet proof" body armor to boat hulls. It is also used to make extremely strong, small diameter rope. Kevlar, like Nomex, does not melt, but chars at high temperature. The char point for Kevlar is about 800° F. (427° C.). On the low temperature end, it has been shown to withstand -320° F. (-196° C.) with almost no embrittlement or strength loss.

Water has little effect on the strength of Kevlar, and its chemical resistance is quite good, with the exception of strong mineral acids and alkalies. It has poor ultraviolet resistance, unless coated with a protective resin. Kevlar fibers are very inelastic, stretching only about 3% before breaking. This makes for ropes which do not absorb energy well, and which are weakened 40% to 60% by knots. Kevlar ropes should only be attached to anchors which can withstand heavy shock loading; and are best used with large descenders which do not bend ropes sharply. Although Kevlar rope will not melt during a fast rappel, it may experience strength loss due to internal self-abrasion of fibers. It should be rappelled on smoothly, with caution.

SIZE

Ropes of almost thread-like size have been used for rappelling, but in ordinary – non-emergency – situations, you shouldn't rappel on anything which might break during normal use. In certain situations where equipment must be carried for long distances, 9mm rope is used extensively in order to reduce weight. However, this is not to say that it is desirable.

Normally, the smallest diameter rope for rappelling should be 7/16 inch (11mm). Most rappelling is done on this size. Half inch (12.5mm) rope is coming more into common use, and except for its weight, is an excellent choice. In situations where safety isn't compromised by weight and clumsiness, ropes of even larger diameter may be useful. Not only are these larger ropes stronger and more resistant to cutting and abrasion, but its far easier to get a scared, untrained person to rappel on a 16mm rope than a 11mm rope.

DYNAMICS

All standard rappel ropes will support loads of several thousand pounds when new. This is called static strength and is an almost useless gauge of rope safety. Rappel loads are rarely static (not moving, or moving slowly). Most involve large amounts of weight sliding rapidly down a rope and being stopped quickly. These are called dynamic loads and represent tremendous amounts of energy. A 7/16 inch wire rope may support from 3,000 to 21,000 lbs., depending upon its construction; but stretch only 3% at breaking. Because of this lack of elasticity, it may be broken by a sudden impact force representing only 0.5% of its breaking strength. Force of this magnitude can be generated by a man falling only a few feet. Because of the lack of stretch, or elasticity, rappelling fast and stopping suddenly could easily snap such rigid line or pull out an anchor. Even if the rope or anchor didn't break, the rappeller would probably be killed from the impact forces generated at the rappel harness.

Wire rope has little ability to absorb large dynamic (extremely quick) loads. Natural fibers and many synthetics don't absorb loads well either; as pointed out in a prior section. Nylon fibers, when bundled properly, can absorb impact loads more reliably than any other presently used rope material.

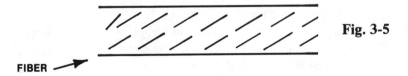

Fig. 3-5

FIBER

In a fiber of nylon, molecules run at angles to the axis of the fiber (Fig. 3-5). Under load, as the fiber is stretched, these molecules twist into a more linear direction, absorbing energy in the process. When the load is removed, they tend to go back to their original positions. If the load was light, the molecules return to their original orientation. If the load was great, permanent deformation results. After many repeated loads, the molecules' positions change greatly and the elasticity and strength of the fiber – and rope – is essentially lost. The energy resulting from this stretching is converted mainly to heat, because of friction between molecules and between rope fibers.

Dynamic climbing ropes will stretch over 40% before breaking. This elasticity is good for absorbing shock loads, but not good in most rappel situations. Static rappel ropes stretch less than 20% at break, and their breaking strength, when new, is specified by the manufacturer. However, these ratings can be very misleading.

Some manufacturers rate their ropes by the lowest breaking strength encountered in testing. This means that their ropes will hold at least as much as is claimed. Some manufacturers use the highest measured strength; others use the average breaking strength. Without knowing the rating system used, and the method of testing, it is hard to determine the actual capability of a rope. Fortunately, most rappel ropes are strong and reliable, and knowing the exact strength is normally of little importance.

What is important is knowing what percent of the breaking strength can be safely used. This is called the working load. The working load recommended for industrial static ropes (which are not designed to support life) is normally about 11% of their breaking strength. When a rope is used for rappelling, a working load of 7% or less (a safety factor of 15:1) is recommended. However, a hard rappel on a low stretch rope might generate forces which exceed the recommended working load, and in some cases, the breaking strength of the rope.

If static strength is not a absolute indicator of rope safety, then the primary rope strength of importance to the rappeller is dynamic strength – or the ability to absorb the energy produced by sudden loads. A rope must stretch like a rubber band, instead of breaking, and cushion the load on the rappeller and the anchor. Stretch, of course, is only desirable at high loads. The perfect rappel rope would have no stretch at all at bodyweight. There are presently no standards for dynamic capacity of static rappel ropes; but there is such a standard for dynamic climbing ropes. Every rappeller should know something about this test system, for it shows how ropes respond to sudden loads.

The International Union of Alpine Associations (U.I.A.A.) devised a test for judging dynamic rope strength. A weight of 176 pounds (80 kg.) is tied to the end of a 9.2 foot (2.8m) rope and dropped from a height of 16.4 feet (5m); see Fig. 3-6. The rope under test is attached to a rigid anchor halfway up the drop and run thru a 10mm carabiner or 10mm metal plate. The weight is dropped and the forces on the rope and weight are measured. The speed of the weight, just before it applies a load to the rope, is 32.81 feet per second (10m/sec.); or the acceleration speed of gravity. Because the length of the fall is almost twice the length of the rope, this test represents a fall factor of 1.78.

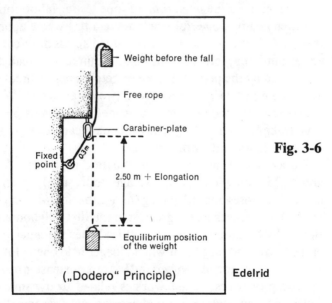

Fig. 3-6

Weight before the fall

Free rope

Carabiner-plate

Fixed point

0.3 m

2.50 m + Elongation

Equilibrium position of the weight

("Dodero" Principle)   Edelrid

To pass the test, an 11mm climbing rope must withstand five falls without breaking, and apply a restraint force (impact force) of no more than 2,646 pounds (1,200 kg.) to the weight on the end of the rope. Because the weight reaches the acceleration speed of gravity (9.8m/sec.), the impact force on the weight and rope is at maximum, and would be no greater even if the drop height was miles instead of feet.

This U.I.A.A. test represents a "worst case" situation and puts maximum load on a rope. Rappel falls and drops are normally no longer than the length of the rope; or a fall factor of 1 or less. Impact forces in the system are therefore much lower than with greater fall factors. However, static ropes do not absorb energy nearly as well as dynamic climbing ropes, so caution should be used when rappelling fast and stopping suddenly. Most quality static ropes will take perhaps two U.I.A.A. class falls, but the impact forces generated would kill or injure a climber. This is why static ropes must not be used in most belay situations.

The energy absorption capacity of a rope is limited by rope construction, moisture, temperature and the radius of the edge over which the rope must run. In rope testing laboratories, ropes are stretched by powerful machines in a test which approximates the U.I.A.A. setup. The ropes are pulled across different diameter edges until they break. The load required to break the rope, multiplied by the percent the rope stretches before breaking, gives a figure known as "working capacity over an edge" (WCOE). It is normally expressed in metric as metre-kilograms per meter of rope (mkg/m), and along with other data can be used to evaluate the strength and safety of a rope.

An average 11mm diameter dynamic rope has a WCOE of about 250 mkg/m. The average rope, used under average conditions, loses about 1.6 mkg/m of capacity for every hour it is used. The strength loss is greatest in the first 100 hours, and then slows a bit for the rest of the rope's life. When the capacity of a rope drops below 160 mkg/m, it will no longer hold one U.I.A.A. fall.

Arova Lenzburg of Switzerland has done extensive research into the aging of ropes. Their work is related to dynamic climbing ropes; but since static ropes undergo some of the same stresses, the findings relate to them also. They found that Mammut ropes, used under average conditions of climbing, lost 0.4 mkg/m dynamic and 0.09% static strength per hour of use. This would indicate that there is a definite number of hours which any rope can be used before its strength, static or dynamic, decreases to an unsafe level. The figures given are only an average for certain Mammut dynamic ropes. Hard use can cause a rope to lose strength at much greater rates, and different brands and different type ropes will age at very different speeds.

As stated earlier, the U.I.A.A. test specifies that the rope be run over a 10mm diameter bar or edge. A rope which will hold seven U.I.A.A. falls in this test may hold only four falls when run over a 4mm edge, and no falls when run over a 1mm edge. What this means to a rappeller is that a rope must never be run over a sharp edge. Always pad sharp edges, or use rope protectors (see Chapters 7 and 11).

Nylon and certain other synthetics can lose from 10% to 25% of their strength when saturated with water. All ropes, even those which do not absorb water, lose some dynamic strength when wet, because moisture between fibers is squeezed out during a sudden load. This "squeezing out" process makes the rope stiffer and prevents it from stretching to its full potential; thereby decreasing the amount of energy which can be absorbed. Chemicals can be put on rope fibers to prevent them from absorbing excessive moisture; this can reduce the strength loss to less than 10%. These treated ropes are referred to as "waterproof" or "dry."

Besides losing strength during use, ropes do all manner of things. For instance, as kernmantel ropes are pulled over rough surfaces, fibers are pulled out to the side and the ropes become shorter. Along with becoming shorter, they become more flexible. This is generally bad because it increases knotting difficulty and the need for friction in the rappel system. The stiffer the rope, the less friction is needed in the descender to hold a given amount of weight. Some ropes are manufactured stiffer than others; these are best for rappelling.

It has been suggested that boiling ropes, particularly nylon ropes, can renew the strength of the fibers after a rope has been used for some time. The "new feel" of boiled ropes is actually a stiffening which results from fiber shrinkage during boiling. Realignment of molecules, for renewed strength, has not occured, and the rope can actually be weaker than before the boiling.

Rope fatigue resulting from use and age cannot be reversed by any known process. Do not boil ropes – and do not rappel on one which has been boiled.

There is also a belief around that kernmantel rope remains strong enough to be safe as long as its kern is undamaged. This is totally incorrect! Although mantel fibers lose strength the quickest, much testing has shown that the kern also loses capacity during each use of the rope. The fact that the mantel looks perfect means little in terms of WCOE.

LIFE SPAN

The life of a rope is shortened primarily by five things: abrasion, heat, pollution, sunlight and load. Abrasion damage occurs anytime the fibers of a rope are cut. This usually occurs rather gently, from the rope being pulled over rough surfaces like rock or bark, and results in fuzz on the rope surface. Thin fuzz, on a new rope, means little loss of strength. In fact, it helps protect the fibers beneath from abrasion and slows the aging process. However, it might be wise to retire a rope when it gets extremely fuzzy, because each cut fiber means strength loss. Heavy abrasion occurs when the rope is pressed in contact with sharp edges – like sharp cliff edges or knife blades. When this occurs, very long fuzz appears and it is definitely time to retire the rope. This is the reason to watch for sharp edges. Rope, with a load on it, cuts very easily.

Certain types of abrasion do not produce rope fuzz, but may do great damage. For instance, when you rappel very fast, the heat generated by the friction may become great enough to melt the surface of the rope. No fuzz will be produced, the rope will feel rough and perhaps hard, and the strength loss can be very great. A rappeller once went down a 1,200-foot deep pit so quickly that when he stopped, near the bottom, his rope melted apart!

Pollution, either in the form of air pollution or chemical contamination, can greatly affect rope life. As was mentioned earlier, strong acids, alkalis, and even some oils can weaken rope materials. Household bleach, lard, battery acid, lye, and even acid rain have reportedly damaged ropes. Just one contact with such substances could cause a rope to become unsafe.

Ropes made of synthetic material won't rot, but they should always be stored in a dry area. Ropes made of natural materials, such as hemp or manila, must never be stored wet, because they will quickly mildew and rot.

Temperatures over 140° F. (60° C.) will permanently affect the strength of some ropes. Others can take higher temperatures. The temperature quoted is for one brand rope made of Type 6 nylon. Low temperatures can also affect strength and elasticity. Although nylon can normally take very low temperatures, as indicated earlier, some tests have shown that nylon ropes, which were placed in water and then frozen, lost up to 40% of their strength at temperatures of -104° F. (-40° C.).

The ultraviolet portion of sunlight can damage most ropes. It has been shown that Type 66 nylon loses over 90% of its strength after one year of exposure to tropical sunlight. In the less harsh sunlight of northern regions, it may lose only 30% after one year. Type 6 nylon loses strength even faster. Polyester, particularly DuPont Dacron, is not affected by ultraviolet as much as nylon and certain other materials; but it is affected.

Because light has to penetrate fibers to be destructive, thick ropes (with more fibers) last longer than thin ones. If a 5mm rope loses 50% of its strength, an 11mm rope may lose only 33% under the same conditions. The color of a rope is also a factor. Some dyes and colors block the destructive ultraviolet better than others, with dark colors usually providing more protection than light colors. Although light is damaging to most ropes, it can generally be disregarded – as long as the rope is only exposed during use. Always store ropes in dark areas when not in use.

The amount of force to which a rope has been subjected (load) also affects its life. Each time a rope is stressed by a heavy load or impact it gets weaker. Eventually, it has little strength left. A number of hard rappels, one U.I.A.A. class fall, or towing one car should bring on the retirement of a rappel rope.

There is no way of saying exactly when a rope is too old or worn to be used; but we do know, from data developed by Arova Lenzburg, that ropes lose a certain amount of strength each hour they are used. For instance, if a certain brand of dynamic rope loses 1 mkg/m strength for every hour of use, and when new had a capacity of 360mkg/m, then it can possibly be used 200 hours before retirement. 360mkg/m (new strength) minus 160mkg/m (minimum allowable strength) divided by 1mkg/m (strength loss per hour) = 200. Wisdom might suggest retiring the rope even sooner because other factors, such as age, may lower its strength. A U.I.A.A. test was performed on some nylon ropes carefully stored for 20 years. They broke on the first drop! Hard use can cause a rope to become unsafe very quickly, various brands have different aging rates, and static rappel ropes age differently than dynamic ropes. The computation above is given as an example – not a recommendation of how long you can safely rappel on your rope!

A log should be kept on the use of every rappel rope. This log should contain: the address of the rope's manufacturer, date and place of purchase, the date and number of hours of each use, and info on nature of use. By comparing the total number of hours of use with the estimated loss of strength per hour, it becomes possible, to some extent, to keep track of the rope's remaining strength. When keeping record, don't be too conservative. If a rope was used 4.6 hours, write 5 in your log.

A good rope is expensive, and it is hateful to have to retire it; especially if it still looks good. Just bear in mind that your life is worth far more than the cost of a new rope!

CARE

If you use your rope intelligently and care for it properly, it can give a long and useful life. Never let a rope run over sharp edges; always protect it (see Chapters 7 and 11). If possible, do not let a rope down or pull it across a rough surface. This will break the tiny outer fibers. Use a protector, or lay something over the rough edge before hoisting the rope. Dirt and grit can get into the interior of a rope and abrade fibers. For this reason, try to keep your rope clean. Never walk on it, and if possible, keep it out of the mud.

Ropes can be safely washed in lukewarm water which has a little mild detergent added. Make sure the detergent has no lye, bleach, or oil in it. One of the products made for washing rope or delicate fabrics should do well. Keep the water temperature low, like below 110° F. (43° C.). Hot water won't clean any better than lukewarm, and might damage certain ropes. If a rope is heavily caked with mud, it is helpful to first run it thru a SMC rope washer (Fig. 3-7). Just hook the washer to a water hose, and pull the rope thru the unit. Once the worst dirt is removed, the rope should be washed in a machine set on the gentle cycle. After washing, rinse the rope thoroughly, and loosely drape it over a shaded clothes line till dry. Do not kink or tightly stretch the rope during the drying process.

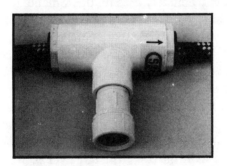

Fig. 3-7

When your rope is dry, you may find that it has a weird, powdery feel. That's a special oil which was put on the fibers during manufacture of the rope, so that they would feed smoothly thru the weaving machines. The lubrication on the fibers also increases rope strength by decreasing internal fiber abrasion.

Store ropes in dark, cool, dry areas away from all chemicals, including moth balls. Store them in cloth or polyethylene bags. Never hang a rope for long periods over a small peg or nail, and don't leave any kinks or knots in it during storage.

MARKING THE MIDDLE

It is very important to know the precise middle of your rope, so that when rappelling double-line the rope ends will be at the same height. Most static rappel ropes are sold in continuous lengths and the middle must be marked by the purchaser. All U.I.A.A. approved climbing ropes have the middle point marked with cloth tape; a colored area; or, in the case of bi-color ropes, the color change takes place at the middle of the rope. Cloth tape tends to come loose after a number of rappels, and it should be replaced with more tape. Just be sure the adhesive on the tape is a type which will not soften rope fibers.

Avoid using any type of paint to mark the middle; except perhaps cloth paint designed to be used with nylon or whatever your rope is made of. A cold solution of Rit dye might also be used. Make sure it is cold – never put your rope in hot liquids. Once again, the best advice is to contact the manufacturer about things which can be safely used on the rope.

FIXING ROPE ENDS

Rope ends must not be allowed to fray, lest some strength be lost near the ends. This is most important with laid rope, but is also important with kernmantel. The classic method of securing the ends is whipping the rope with thread (Fig. 3-8). This is still a valuable way for ropes of natural materials, but there are better methods for most synthetics.

Fig. 3-8

Fig. 3-9

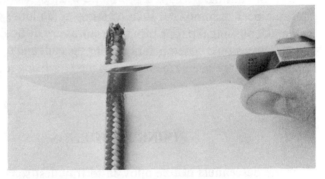

Fig. 3-10

Fig. 3-11

The preferred method for most synthetic ropes is melting the ends. However, simply cutting the rope with a hot wire, or holding the end in a flame, does not always do a good job. This usually leaves an end which is rough or which is larger in diameter that the rope itself. This can cause the rope to get hung when pulling it thru tight spots.

The best way to fix the ends is to gently melt the outer fibers with a torch (Fig. 3-9), cut thru this seared area with a sharp knife (Fig. 3-10), and then sear the end (Fig. 3-11). Form the end as you melt it, so that it comes out smooth and smaller in diameter that the rope. This is also the best method to use when cutting a rope into sections.

If the rope is natural, or one of the synthetics which chars instead of melting, perhaps a good tape wrap or whipping might suffice. Bear in mind that such methods come loose easily, and can make pushing the rope thru protectors very difficult. Perhaps a better way would be temporarily taping the ends and soaking them in epoxy resin; or one of the resins designed for fixing rope ends.

UNKINKING

Some descenders cause ropes to twist and kink badly, and it gets worse on each succeeding rappel. Laid ropes kink no matter what descender you use; although some are worse than others. These twists can be eliminated to a great extent by hanging the rope from a high anchor, so that it is free to untwist. Shaking it for several minutes will expedite matters. Since ropes sometimes kink during machine washing, a good time to do the unkinking is after the wash job. Just let the rope dry first.

ROPE HANDLING

Handling covers how a rope is coiled for storage and how it is released or payed out for use. Different situations will require different methods.

The simplest method of coiling is the basic coil, shown in Fig. 3-12. It is excellent for long storage of ropes, because it puts no sharp bends in the line. If you are coiling laid rope, remember that right-laid should be coiled clockwise, and left-laid counterclockwise. Most kernmantel and braided rope can be coiled as you like. Uncoiling and letting down the rope is basically the reverse of the coiling procedure. Don't try tossing a large coil over the edge and expect it to uncoil without tangles. Uncoil it on top and let it down by hand.

Fig. 3-12

The method shown in Fig. 3-13 is called the Swiss Coil. Coiling is started by grasping the rope at its center and winding it into a short, flat coil. The two ends are wound several times around the top, and then taken thru the top loop. The ends should be left about 6 ft. (2m) in length.

Fig. 3-13

The rope is payed out by doing the procedure in reverse. Like the basic coil, it should not be tossed over the edge. The virtue of the Swiss Coil is that it can be carried like a backpack (Fig. 3-14). Simply take the ends over each shoulder; around and behind the coil; and back around the waist, where they are tied. This is a good method for carrying the rope while climbing.

Fig. 3-14

The Log Coil, as shown in Fig. 3-15, is a good method for quickly uncoiling rope without tangles. The log is simply tossed over the edge, and the rope unwinds from around it. Its main problem is that of killing someone on the ground (Fig. 3-16). Also, as the log falls it tends to do weird gyrations. This can cause the rope to become entangled in trees and etc.

The Monkey Chain, or Chain Coil, is an excellent way to have quick, tangle-free pay outs. It is simply a series of slip knots (Fig. 3-17), with the end taken thru the last loop to form a half hitch. To pay the line out, simply undo the last loop, hold on to

Fig. 3-15

Fig. 3-16

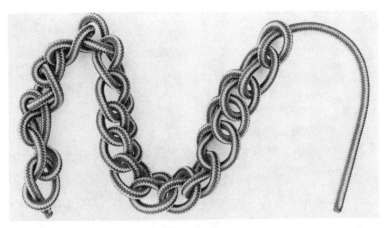

Fig. 3-17

Fig. 3-18

the end and toss the chain over. If made correctly, it will pay out smoothly and do fairly well even when falling thru trees. For more reliability in thick growth, the Monkey Chain can be placed in a cloth bag before it is tossed down. The slip knots should not be tied tight, and because of the sharp bends in the rope, this method should not be used for long periods of storage.

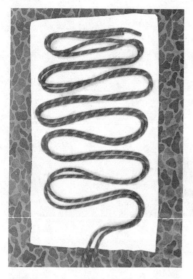 

Fig. 3-19   Fig. 3-20

The slip knots of the Monkey Chain can also be tied around a basic coil (Fig. 3-18). Start out with a few turns and then form the rest of the rope into slip knots around this coil. This method pays out better than a Monkey Chain, when falling thru trees.

The Fake/Bag is a way to have smooth pay-outs and long term storage in one method. As seen in Fig. 3-19, the rope is faked inside of a cloth bag. This should be done smoothly, without kinks. Once the rope is in, the bag is rolled into a tight bundle (Fig. 3-20). The rope can also be simply stuffed into a rope bag, although the pay-outs may not be as reliable. To uncoil the rope, the end is held while the bag is tossed over the edge.

In many situations, particularly tactical ones, it is not desirable to pay out the rope before rappelling. In these cases a rope bag can be strapped to the rappeller's waist or leg, and the rope payed from the bag during rappel. If a bag isn't available, the rope can be faked inside of a pants leg. Tie a cord around the bottom of the pants leg to keep the rope from falling out. When rappelling in this manner, it might prove safest to tie the tail of the rope to your harness, lest the rope not reach the bottom of the drop.

BODY RAPPELS

Only two elements are absolutely necessary for rappelling. One is a rope and the other is the rappeller. The rappeller can use his own body to provide the friction necessary to slow the descent. These type rappels are called body rappels, but because of their pain and danger, are seldom used nowadays. However, every rappeller should be experienced at using them – in an emergency situation they may provide the only way down. Until experience is gained, these rappels should be practiced on very short drops and only with a belay (see Chapter 10).

The most well-known method of body rappelling is called the Dulfer or Dulfersitz. As can be seen in Fig. 4-1, the rope is run between the legs, around the hip, across the chest, over the opposite shoulder, and across the back to the braking hand. Descent speed is controlled primarily by how tightly the rope is gripped by the braking hand. Extra friction can be gained by wrapping the rope one or more times around the braking arm, as shown in Fig. 4-2. It is important to keep this arm fairly rigid, and well away from the body. If it is held too close, and one has to brake suddenly, it could be pulled up behind the back. Be very careful when using an arm wrap for extra friction.

It cannot be stressed too much that body rappels are very dangerous and can be extremely painful! Where the rope runs across the shoulder, pressure on local nerves can cause a loss of muscle control – and even paralysis of the arm. A girl once had to do a 300-foot Dulfer to get off a mountain. It cost her much time in the hospital. A ranger had to do a long Dulfer to rescue a climber. The rope cut thru his clothing, thru his flesh, and before the rappel was over, was rubbing against bone.

To reduce the pain and increase the safety of body rappels, it is important to wear gloves, use double rope, and rappel slowly. It also helps tremendously to pad those areas where the rope contacts the body, particularly the shoulder, hip, and groin. This can be done by sewing leather pads on your clothing, or stuffing cloth, leaves, bark, etc. inside your clothing at these points. In an emergency, removing your socks and placing one on your shoulder and the other under the hip, could spell the difference between a safe rappel and a deadly one.

Fig. 4-1

Fig. 4-2

To do a Dulfer, place the rope around your body as shown (or opposite if you are left-handed) and carefully back over the edge, keeping your weight on the soles of your feet, as discussed in Chapter 11. Dulfers are well suited for free drops (where your body doesn't touch the rock) because the rope makes first contact with your body at its approximate center of balance. Once you are over the edge, proceed down the rope, traveling slowly enough to keep the rope from burning through anything. Keep your balance hand in place, and remember that it is only for balance – do not try to hold weight with it.

A variation of the Dulfer is called the Geneva (Fig. 4-3). The rope is passed across the hip and then wrapped around the forearm, instead of going across the shoulder. Genevas are generally less painful than Dulfers; but they don't work well if there is a great amount of rope weight below the rappeller.

Fig. 4-3

Another variation uses a seat harness and carabiner to make things safer and less painful (Fig. 4-4). The rope is run thru the carabiner, over the shoulder, and down to the braking hand. This is one of those techniques which is utterly ridiculous! If you have a carabiner available, why not use a Carabiner Wrap for braking, and avoid the pain altogether? For more information see Chapter 6, "Carabiner Wrap."

Fig. 4-4

The Tarzan method of rappelling is shown in Fig. 4-5. Friction is developed by wrapping the rope around one leg and sliding down. The great drawback to this method is difficulty of control. Best to leave this trick to those who wear loincloths.

An excellent body rappel method is the Arm Wrap or Hasty Rappel. As can be seen in Fig. 4-6, the rope is wrapped around one arm, goes under the armpit, across the back, and is wrapped around the braking arm (the one pointing toward where you intend to go). It is imperative that the rope run under the armpits. If it runs over the shoulders it might saw thru the old neck – this

Fig. 4-5

Fig. 4-6

would be painful. Although normally used for walking down steep grades, Arm Wraps can be used for short vertical drops (Fig. 4-7). However, this can be very painful and dangerous. Besides rope burns, there is great danger of your balance arm being suddenly jerked upward as you drop over the edge. This could easily pull a shoulder out of socket and ruin your whole day! Vertical drops must be done with extreme caution.

Fig. 4-7

It is extremely important to wear long, heavy sleeves and gloves when doing Arm Wraps. Without proper protection, even ceiling height rappels – done slowly – can result in very painful rope burns. In a survival situation, because of stress and infection, even simple rope burns can lead to death. This advice applies to any kind of rappelling in which the rope might make contact with bare skin.

It is also very important to make sure the edge you rappel from is sturdy, and cannot break as you go over. Even short falls during a body rappel can be disastrous.

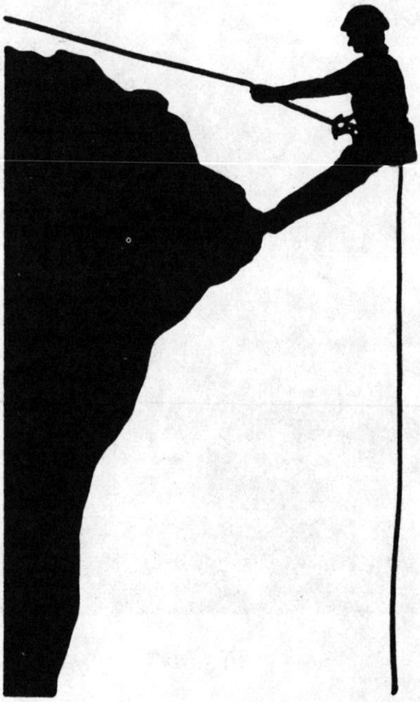

HARNESS

If you only do body rappels, you'll never need a harness. But if you're like most folks and prefer rappelling with some measure of comfort and safety, you will definitely have to have a harness. The questions are: What type is best? Which should you buy? Or should you make your own?

If you're a mountaineer, who seldom rappels, your needs can probably be met by a simple Swiss Seat made of narrow webbing. If you're a climber or caver, or in a rescue squad, you may have occasion to hang in a harness for long periods. Unless your pain threshold is very high, you'd better have a comfortable harness.

Comfort is also related to safety. A comfortable harness will allow longer hang times before physical and psychological problems occur. This is very important when raising or lowering injured people. Long periods of hanging can result in extreme pain, numbness, blood pooling in the legs, blood pressure changes – and fatal mistakes. Depending upon the harness used, comfortable hang times range from a few seconds to almost an hour.

The simplest harness available is the Pompier, or Fireman's Waist Belt (Fig. 5-1). Its been around for a long time, but so has acne. Neither is desirable! A belt around the waist throws the rappeller completely out of balance; he becomes bottom-heavy, which makes getting over edges hazardous. If the belt doesn't fit right, it can slide up under his arms, or allow him to fall out.

A Waist Belt is also very uncomfortable, and under certain circumstances can be deadly. A man hanging in a narrow waist loop has only minutes to live! Luckily, the Pompier is made of wide material, and considering the length of time one is usually in it, there are few problems. However, long hang times can be fatal; and taking a fall could result in a broken back. Never use a Waist Belt except in an emergency – and if you do have to use one for a long period, try sitting in it, instead of putting it around your waist. Just be super careful about falling out.

The next simplest harness is called a Figure-8 (Fig. 5-2). Just take a short loop of webbing (or rope, if you can stand the pain), and slip it over your legs and up to the crotch. Clip onto both strands, between your legs, with a carabiner. Make sure it fits snugly. If the harness is too loose, take it off and twist

Fig. 5-1

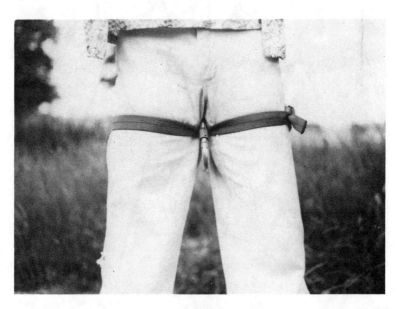

Fig. 5-2

it several times in the middle until the slack is taken up. The reason for the snug fit is because during the rappel you may accidentally turn upside down. If the harness fits tight, you may not fall out. The main problem with a Figure-8 is that you are off balance when sitting in it, which increases your chances of turning upside down. Once upside down, there is a good chance of falling out. Never use a Figure-8 except in an emergency, when you don't have enough webbing to make a Swiss Seat.

With a longer piece of webbing you can have a Swiss Diaper Seat. Take a length of webbing, about three times your hip measurement, and form it into a loop by tying (see Chapter 8), sewing (see latter part of this chapter), or locking with a secure buckle. Hook this loop on an anchor and jump up and down in it, to test its strength.

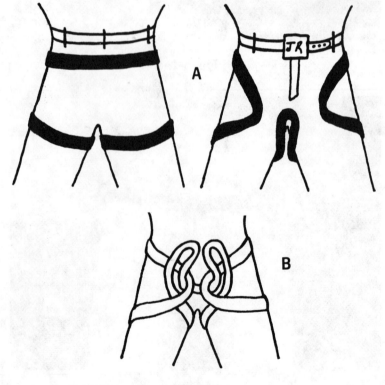

Fig. 5-3

To put this harness on, put it behind your back and bring part of the loop around your left side, part around your right, and then pull a loop up between your legs (Fig. 5-3A). Clip the three loops together with a carabiner – going from side to middle to side. This harness should fit snugly, but not tight. If its too loose, it can be tightened by twisting one of the loops the carabiner hooks onto. Fig. 5-3B shows another method which fits more securely on some. The loop from between the legs is brought around the two upper loops, forming two extra loops. The carabiner is then clipped in to both these loops. Either way, a properly fitted Swiss Seat is so secure that its almost impossible to fall out.

When making a harness, use only strong, high quality webbing, such as nylon or polyester. Never use natural materials like cotton or leather. They rot, or can easily break during a fast rappel stop. Don't use the weaker synthetics either.

If you're in a rescue squad, you might consider sewing a harness into your pants. When needed, it can be accessed thru your fly, or thru the top of the pants. Before sewing, position it so that it will give proper balance; and sew it in with Nomex thread if you'll be doing fire rescue. If you buy a commercial harness, some suppliers will sew it into your pants for you, for a small fee.

Fig. 5-4 shows the harness used at the Army Air Assault School. It is made from about 12 ft. of ½ inch nylon rope. To tie this harness, place the middle of the rope behind your back; bring the ends around front; and loop them twice, forming a snug Double Overhand knot around your waist (Fig. 5-5). From here, take the ends between your legs, up across the hips (Fig. 5-6), and tightly pull them up thru the waist loop (Fig. 5-7). Tie the ends together on your left side with a Square knot (Fig. 5-8), secure them with Half Hitches, and stuff them into your left pocket.

This isn't the most comfortable harness in the world, but it is safe for short periods. Just make sure the Square knot is tied correctly. The main virtue of this harness is that the carabiner can be attached in front; on the side; or in the waist loop in back, for doing Australian Rappels. Normally, the harness carabiner is clipped onto both the Double Overhand knot and the rope above it (Fig. 5-4), and is used as the descender.

Fig. 5-4

Fig. 5-5

Fig. 5-6

Fig. 5-7

Fig. 5-8

If you don't wish to make your own harness, you can buy one. They are available in all shapes, sizes, and prices (Figs. 5-9A and 5-9B), and many will meet the various national and international safety standards. Before you buy, it might be wise to talk to people who have used the type you are interested in. Ask if the harness causes any pain, or if they've taken any long falls while wearing it. Before buying any harness, make sure there is no way of falling out, and note the difficulty of getting into it. It's best to have a harness which goes on quickly and easily – it reduces the chance of a fatal mistake.

When purchasing a harness for rappelling, make sure that the carabiner attachment point is near your body's center of gravity (center of balance). On most folks this is about midway between crotch and navel. If the attachment point is much higher or lower than this, there is increased chance of losing your balance when backing over an edge. In a properly fitted harness, you should (with your hands on your hips) be able to stretch out horizontally without either end of your body quickly rotating downward.

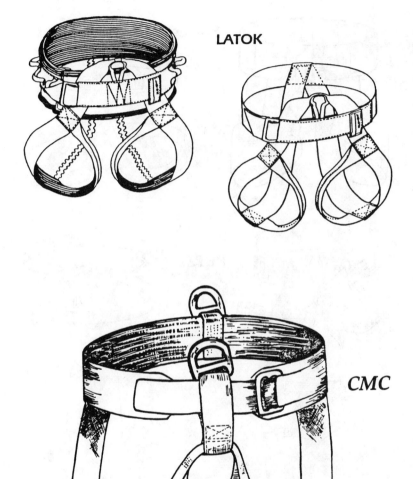

LATOK

CMC

Fig. 5-9A

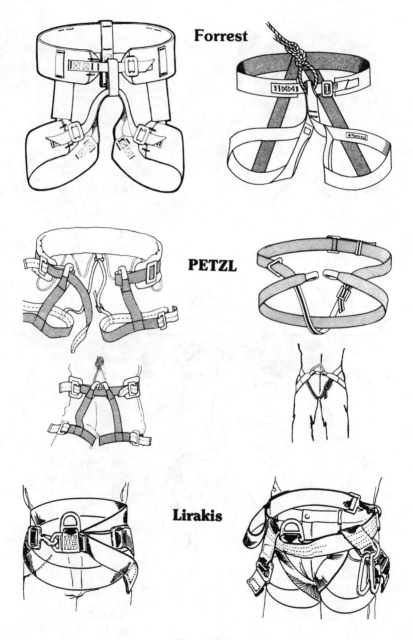

Forrest

PETZL

Lirakis

Fig. 5-9B

In most climbing and hoisting situations, it is important that the body be kept in an upright position, a few degrees off vertical. This can be accomplished by moving the attachment point higher, by means of a chest harness. The simplest chest harness consists of a webbing sling twisted into an "8" shape, with the center of the "8" in back, and an arm thru each loop (Figs. 5-10 & 5-11). The carabiner is clipped onto each of the arm loops in front. When necessary, the loops can be twisted so as to get a snug, but comfortable fit.

A chest harness must never be used to hold full body weight, because a person can slide out of it. It is only for balance, and should be connected to a seat harness with a short sling. This converts the whole affair into a body harness. Never hoist anyone with only a chest harness. Injury or death could be the result. Be cautious when rappelling with a descender connected at chest level, because this throws you completely out of balance. Always attach descenders at seat level if possible. Fig. 5-12 shows several commercial chest and body harnesses.

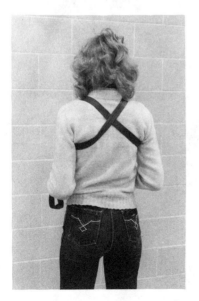

Fig. 5-10   **Fig. 5-11**

When rigging a body harness, do not merely tie the rope to the seat harness, and then run it thru the chest harness carabiner. In a fall, such an arrangement can pull the chest towards the crotch, causing severe back and internal injuries. Connect the seat and chest harnesses together with a strong sling, and use the chest harness as the suspension point.

Although the body should normally be kept upright, such a position can be lethal if one is unconscious, or is in pain or shock. Under these conditions, the blood flow to the brain is reduced, and death may follow if the person is not quickly placed in a horizontal position. For greatest safety, harnesses should be rigged so that a person can be easily released from the vertical and allowed to hang in a more supine position.

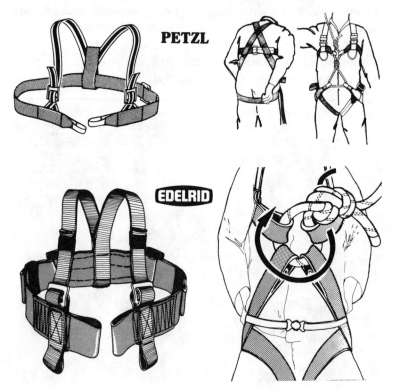

Fig. 5-12

Like rope, harnesses do not last forever. Their life is shortened by abrasion, light, air pollution, heat, load, etc. In one test, webbing was shown to lose 30% of its strength after only 100 days of use. Edelrid recommends that harnesses made of webbing be used a maximum of five years, and that they be retired earlier if threads or webbing are cut, or excessive wear is evident. Harnesses should also be retired after holding one hard fall. They must be stored and cared for in the same manner as rope, because your life depends upon them too.

As stated earlier, the amount of difficulty of getting into a harness plays a great role in safety. Particularly in night rescue operations, simplicity and speed are of tremendous importance. If you require a harness which is very comfortable, simple, low cost, and fast, you might consider making a Super Swiss Seat (Fig. 5-13). Here's how: Purchase some three inch wide nylon or polyester webbing – about three lengths of your hip measurement. This webbing should hold at least 8,000 lbs.

Next, purchase three descending rings. They should hold at least 3,600 lbs. If you intend to sew the harness yourself, buy a spool of heavy duty button and carpet thread. This thread is made of polyester or nylon, and is so strong that you probably won't be able to break it by hand. Also, buy a color which will contrast sharply with that of the webbing. This makes spotting cut or damaged threads very easy.

The reason for sewing, instead of tying the ends, is because this webbing is so bulky. A knot would turn out to be awfully big. Also, a knot is the weakest point on webbing; whereas a properly sewn splice is stronger than the webbing itself. Some home sewing machines will handle the heavy thread required, but if yours won't, you can always sew it by hand – or take it to a shoe repair or upholstery shop, where they will sew it for you for a modest fee. Just make sure they use nylon or polyester thread – or Nomex thread, if you do fire rescue.

The first part of this operation is fitting the webbing to your size. Thread the three descending rings onto the webbing, and without twisting it, lap both ends and temporarily secure them with clothes pins or safety pins.

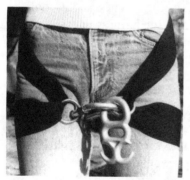

Fig. 5-13

You get into this thing like a regular Swiss Seat, except that you clip onto the rings instead of the webbing. Getting three loops of three inch webbing into a carabiner is quite a job – that's the primary purpose of the rings. They also make this one of the fastest harnesses to put on.

To adjust the size, get into the harness and remove the pins from the lapped ends. Pull on the ends to tighten the harness until it feels very snug, but not tight. Put the pins back in and mark the webbing so you'll know where to sew it. Leave about one foot lapped and cut off the rest.

There are two recommended ways to sew the webbing. The simplest method is to sew back and forth the full length of the lap. By sewing with the "grain" of the webbing, instead of across, the thread sinks down into the fibers. This protects the thread from abrasion. The problem with sewing the full length of the lap is that it makes the lap quite stiff; also the continuous thread runs may loosen easily if they get cut. Perhaps a better method is to sew in "blocks," instead of the whole length of the lap. Sew with the grain and zigzag down the width of the webbing – and make at least four blocks (Fig. 5-14). With this method, the lap isn't stiff and if the thread in one block gets damaged, it doesn't affect the strength of the other blocks.

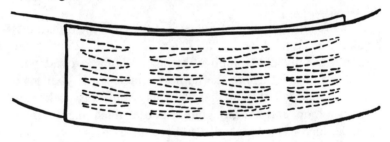

Fig. 5-14

Whichever method you use, be sure and put enough stitches to make a strong joint. Regular lockstitches will hold 5 to 18 lbs. per stitch, depending on the size thread used. So put at least a thousand stitches in your harness. Don't cover the webbing with stitching however; this might weaken it from the needle cutting fibers during sewing. It might be preferable to use polyester thread if you intend to sew by machine. It is less elastic than nylon, and works better in some machines. It is also more resistant to damage by sunlight.

You should test all your equipment by hooking it onto a solid anchor and jumping up and down in an attached sling. Test your sewing in this manner, along with your harness carabiner.

Don't worry too much about the descending rings on this harness breaking; this is possible but not probable. There are three of them, so each has to hold only about 1/3 of the total impact force of a fall or quick rappel stop. If each ring will hold 3,600 lbs. (1,633 kg.) when new, then it would probably take a force of about 10,800 lbs. (4,899 kg.) to cause any breakage. Your body would be torn apart long before that!

When stressed excessively, most rings and buckles will bend or crack before they experience catastrophic failure. As you do regular inspection of your equipment, you should look for these signs. If you only use quality equipment, and treat it with respect, you will probably never see such damage. Far more concern should be directed towards making sure that rings, buckles, and knots on harnesses are secured properly during use. A buckle which tests at 12,000 lbs. is of no value if rigged improperly!

If you're worried about a Swiss Seat (or any type of harness) breaking or coming loose, you can add a bit of safety to the system. Simply put a loop of strong webbing around your waist, and connect it into the harness carabiner. If your main harness should fail, this auxiliary one may hold. Just make sure that it's a loop, and not any kind of noose which might tighten up under load. This loop will also keep the harness from sliding down as you walk, and can be used to move the descender attachment point more near your body's center of balance.

Like all rappel harnesses, the Super Swiss Seat must enable you to hang in a balanced position. If it makes you feel too top heavy, throw it away and try something else. Never use any piece of equipment which makes you feel uncomfortable.

DESCENDERS

Also called Rappel Devices, or Abseilbremse, descenders are a type of rope brake which enables a rappeller to descend a rope in a controlled manner. He can control his speed, or stop if he wishes. His hands have to hold only a small portion of his total weight. The pain is taken out of rappelling!

A descender called the Snake is shown in the cover photo of this chapter. It is attached to the rope and connected to the rappeller's seat harness. Descent control is possible because there is friction between the rope and the Snake. The amount of friction is determined primarily by how much downward pull is applied to the braking end of the rope (the end nearest the ground). This downward force can be applied by the rappeller, or by someone on the ground. If enough "pull" is applied to the rope, the friction between the rope and Snake will become great enough to hold the weight of the rappeller. If the system is set up properly, the rappeller can control her descent with two fingers. Most descenders work in this fashion.

Heat is produced by the kinetic energy released during the descent. Most of it results from friction between rope fibers; a lesser amount from friction between the rope and the descender. The total amount of this heat depends upon the height of the rappel and the weight of the rappeller. The temperature of it is determined by these factors plus the speed of the rappel. Fast rappels get things hotter than slow rappels. A number of other factors, like air temperature and humidity, also affect maximum temperature. To sum it up, a descender is a device which converts some of the kinetic energy of the descent into heat and then dissipates it into the air. A tiny percentage of this energy is also converted into sound, as the rope rubs against metal.

Descenders could be built which convert the descent energy to electricity, light, or sound. Practicality, however, suggests that conventional methods be followed. Anyone interested in designing descenders would be wise to first work toward improvements in current technology. Improvements in the areas of strength, weight, and heat dissipation capability would be welcomed by most rappellers. When designing, remember that simplicity is usually the key to overall system safety.

Before getting into devices available, let's look at the materials used to build most descenders. These materials fall into three groups: steel, aluminum, and stainless steel. Various steels have been used in descenders for the purposes of strength and economy. Steels exhibit good strength and wear resistance, but are generally undesirable because of weight and rust. Rust is primarily objectionable because of its ill effects on ropes, and the possibility of early strength degradation of the descender.

Aluminum is the material most widely used in descenders and rappel equipment. Not pure aluminum, but any of the aluminum alloys commonly referred to as Dural, or Duralumin. One of the more commonly used alloys is 7075, with a T6 heat-treatment. It contains .4% silicon; .5% iron; .2% copper; .3% manganese; 2.5% magnesium; .25% chromium; 5.5% zinc; .2% titanium; 88.2% aluminum and .15% miscellaneous.

Besides 7075, alloys such as 2024-T6, 5052-H38, and 6061-T6 are also used extensively. 7075-T6 is the strongest, hardest, and most expensive of this group; so descenders made from it are usually the strongest and most abrasion resistant. However, along with excellent strength and long life, is an inherent danger. Devices built of hard alloys like 7075-T6 may break, before bending, if stressed excessively.

The three main advantages of aluminum descenders are low weight, high friction, and great thermal conductivity. 7075 Dural conducts heat three times better than mild steel (1% carbon), and about eight times better than 17-7 PH stainless steel. Friction produced heat is therefore conducted away from the surface of the rope much faster with an aluminum descender. This prolongs rope life by preventing searing of the outer fibers. With any long, fast rappel on conventional synthetic ropes, melting of the rope surface will be observed. In moderation this is not dangerous; but accidents can happen if the heat is not kept down. The story of the rope melting apart, in Chapter 3, is a good example.

The worst factor of aluminum is its softness. Although this provides the descender with excellent friction properties, clean ropes will eventually cut notches in aluminum descenders and dirty ropes really cut fast; possibly ruining a descender on one rappel.

Descender life can be greatly improved by anodizing. This is an electrochemical process which forms a coating of aluminum oxide on the descender. Aluminum oxide is very hard, and as long as it is intact, will protect the underlying metal from corrosion and abrasion. Several types of anodizing are available, but the one called "hard coat" is the most practical for rappel equipment. A high quality "hard coat" is over .0015 inch (.038mm) thick, and will last thru a couple of miles of rappelling on clean rope. However, even as tough as "hard coat" is, it can be worn off with one rappel on a rope heavily coated with sand and grit. Anodizing can be left clear, or can be dyed any color desired. Black anodizing provides maximum heat dissipation, but does not always produce the hardest surface because of the dye's effect on the aluminum oxide. Simple types of anodizing can be done easily in a home workshop.

One other point: Whenever rope runs in contact with bare aluminum, the rope gets blackened. If you have just purchased a beautiful new rope, and want it to stay that way, avoid letting it run across bare aluminum.

When minimum weight isn't of paramount importance (and it usually isn't), and when rappels can be done at reasonable speeds, stainless steel descenders are the most practical. Depending upon the alloy and heat-treatment used, stainless usually makes the strongest, and hardest, and prettiest descenders on the market. Properly hardened stainless descenders will withstand years of rappelling on dirty ropes. They don't normally rust or corrode, and are about the only things available which won't discolor ropes. Besides weight, the disadvantages of stainless, when compared with aluminum, are low thermal conductivity and less friction. When rappelling with a stainless device, you may have to go a little slower than you would with an equivalent sized aluminum device, in order to allow the heat time to dissipate. You may also have to have a little more rope in contact with metal in order to compensate for the "slickness" of the steel.

Stainless has other advantages. All metals have super-tiny cracks called microcracks. These are cracks between the layers of molecules, and are areas of the weakest bonds. These microcracks are not necessarily bad, for without them, metal could not be bent or formed. During bending or stretching, the microcracks enlarge slightly. If the stress continues, microcracks become macrocracks and the metal nears its point of failure or breakage. Large macrocracks can be seen with the unaided eye, and indicate that the metal has little strength left. The presence of microscopic macrocracks does not necessarily mean a descender is hazardous, but simply that the metal is not as strong as it once was. However, moisture and chemicals can enter these cracks and start areas of corrosion, which can lead to failure. High resistance to corrosion is one of the many advantages of stainless steel.

Descenders should be checked regularly for macrocracks. Any cracks observable with the eye mean that the device should no longer be used. Tests such as "Zy-Glow" and X-ray are available at many labs, and it might be wise to have your equipment checked on a regular basis. Bear in mind that equipment can come from the factory with flaws. Always check new equipment closely.

Descenders and climbing equipment are usually manufactured by being machined, stamped, formed, forged, cast, or welded. Machined devices are made from metal plate which has been cut to shape by sawing, milling, drilling, grinding, or etc. Because the molecules in a metal plate all "point" in the same direction, rather like the grain in wood, very strong items can be produced by merely orienting the grain in the direction that strength is most needed (Fig. 6-1). Stamping is a process in which metal plate is cut to shape by something like a "super cookie cutter." It is a much faster and cheaper process than machining, but the strength of the metal can be reduced by the sudden "crunch." Normally, machining produces stronger devices than stamping.

Formed items are made by bending plate or bar stock into the desired shape. This makes for very strong devices because the orientation of the molecules follows the direction of bend, the way the grain in a tree limb follows the shape of the limb (Fig. 6-2). Problems with forming occur when the metal is bent too tightly, causing macrocracks to form in the outer radius of bends. Also, if the inside of bends are very flattened, it indicates strength loss and poor manufacturing technique.

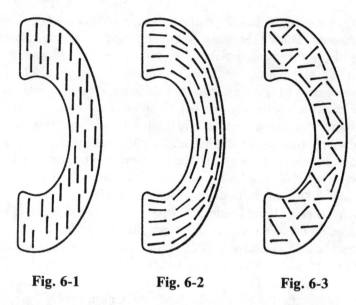

Fig. 6-1   **Fig. 6-2**   **Fig. 6-3**

Forging is a process of "hammering" metal into a mold of the desired shape. Forging, like forming, causes molecules to follow the direction of curves and bends (Fig 6-2), but does so by making the metal "flow" into the proper shape. Forging permits items of almost any shape to be produced in large quantity at low cost. High quality forgings are very strong and reliable, but because of difficulties in manufacturing, most forged devices do not test as strong as machined devices. Low quality forgings can have voids in the metal, poor molecular orientation, and be very weak.

Casting is a process by which items are made by pouring molten metal into molds. Casting permits very low cost production, but is not a preferred method because the molecules in the metal are given no orientation at all (Fig. 6-3). This causes cast items to be brittle and normally low in strength. Castings can also contain bubbles and voids. Cast aluminum devices have been known to break by merely dropping them onto a concrete floor. It might be wise to have cast aluminum descenders and climbing equipment inspected by X-ray, before their first use.

Welding is used in the manufacture of some descenders, and if done correctly poses no safety problems. Weld joints on steel and stainless are usually very strong, and do not weaken the underlying metal, except for certain hardened alloys. Weld joints on aluminum almost always weakens the metal – sometimes to the point of danger. Do not use any aluminum descender which has welded parts, unless you know its tested strength.

Descenders, by the way the rope passes thru them, can be divided into two groups: Those in which the rope runs through in a single plane (Snakes, Pressure Plates, Brake-bar Systems, Racks, Sticht Plates, and etc), and those in which the rope spirals around the device (Carabiner Wraps, Figure-8's, Spools, and etc.).

"Spiraling" type descenders tend to cause the rope below the rappeller to twist and kink, particularly laid rope. This problem occurs with most all descenders, but it is much worse in those devices which spiral the rope. The problem is especially bad because some of these descenders have great difficulty passing the twists and kinks that develop, and can be jammed quite easily.

The problem is further compounded when rappelling on two ropes. The ropes twist together and can really jam the descender. The problem can be minimized by doing short rappels, using kernmantel rope, and rappelling slowly. It can be eliminated by having someone on the ground hold the ropes apart. Many times, however, the rappeller will have to separate his own ropes on the way down, and even remove kinks. It is sometimes advised to place a finger (on the braking hand) between the ropes to separate them. Its best not to do this on a fast rappel – you could ruin a finger. A better way is to place your foot between the ropes (Fig.6-4). This works better and is much less painful. Double-line kinking is why all very long rappels are made on single-line.

Besides causing kinking, "spiraling" descenders may shorten rope life. Anytime a rope is bent around a small diameter rod, the outer fibers are stretched and weakened. Descenders with large diameter sections, or which allow the rope to run fairly straight may improve rope life. They can also affect how much the mantel slips down the kern when rappelling on kernmantel rope.

Descenders can also be classified as variable or nonvariable. Snakes, Racks, Rope Riders, and a few others allow the rappeller to make large changes in the amount of friction produced by the

Fig. 6-4

rappel system – while he is on rappel. This ability is extremely important on long drops, and is potentially helpful on any rappel. It is discussed in detail in Chapter 11. Things like Figure-8's, Sticht Plates, Sky Genies, and etc. are classified nonvariable, because once you are on rappel there is no simple way to radically alter the descender's friction; particularly the minimum level.

The importance of variability is this: Nonvariable descenders cannot be used for long rappels; whereas variable descenders can. Example: You've just rigged a Figure-8 for a 1,200 ft. rappel. You step over the edge, but can't go anywhere. Why? Because there is 138 lbs. of rope hanging below you. Its the same as trying to do a 10 ft. rappel with a 138 lb. man swinging on end of your rope. You can't slide down until you take the weight off the rope below you. If you were using a Rack to do this rappel, you would rig it for minimum friction to start with, so that you could move; then add more friction as needed on the way down.

Things like Figure-8's are useful on short rappels of a few hundred feet or less. You can use them on longer rappels if you are lowering several hundred pounds; or if you are willing to pull yourself down the rope (This is a loser!). To find out if a descender can be used on a long rappel, calculate the weight of the rope necessary, and have someone apply that much pull downward on the braking end of your rope while on a short rappel. Rig everything exactly as you would for the long rappel. If the descender works fine here, it should do well on the long rappel. Just remember that long drops cause descenders to get very hot, and small descenders don't dissipate heat very well.

The matter of how well a descender dissipates heat is one of the major questions to consider before making a purchase. There is no way of telling by looking, but remember that aluminum does better than steel, large (heavy) descenders usually do better than small ones, and dark colors usually dissipate heat faster than light colors. The shape of the device and many other factors enter into this question.

Another factor to consider is whether a descender will pass knots. On very long rappels, it is sometimes necessary to tie two or more ropes together in order to reach the bottom. If you are willing to go to a lot of trouble, you can get around this knot while on rappel, using any descender (see Chapter 13, Connecting Ropes). However, with most descenders the rigging has to be redone as each rappeller descends, so it is usually simpler – and safer – to use a descender which will pass knots. Most Spools and Spindles will pass knots without difficulty, as will most of the larger Figure-8's with Ears. In this chapter, assume that a descender will not pass knots, unless stated otherwise.

One other point: Rappellers have lost their lives because they lost their descenders. A story is told of a fireman, in a burning building, who was attempting to rig for rappel with a Figure-8. In the smoke and stress of the moment, he dropped it. He knew no other way to rappel. He died in the fire!

Some descenders must be removed from the rappel harness in order to hook them to the rope. Others can be rigged while attached to the harness. Obviously, in emergency situations, the latter type is preferable. Speed in rigging is also important, and descenders vary greatly in this respect. No matter which descender you use, be sure that you know other ways to rappel. At the very minimum, learn body rappels and Carabiner Wraps. Your life may someday depend on them!

The rest of this chapter is a catalogue of descenders, purposely arranged in no particular order. Some of them are good and some are not so good. Just because they are illustrated here is no endorsement of quality. Because there are no universal standards for rating descender strength, no breaking strengths will be listed. Harness carabiners are shown attached to many descenders, and most of the ropes shown are 11mm. Use the data presented as a starting point, and then evaluate various devices to see which best meets your needs. Also, remember that price does not always mean quality. Some of the cheapest descenders are very good, while some of the more expensive ones are not so good. Use your head, your life is at stake!

SNAKE

As can be seen in Fig. 6-5, the Snake looks like its namesake. It weighs 15 oz. (425 g.) and is formed from 17-7PH stainless rod, which is heat-treated to a hardness of Rockwell C38 (almost as hard as cheap pocket knife). This level of hardness provides tremendous resistance to abrasion by dirty ropes, thus providing a descender which will withstand a lifetime of rappelling.

Though hard, 17-7PH is a spring steel, and does not tend to fracture and break like most hard steel and aluminum alloys. This characteristic is particularly important in fast, jerky rappels. In fact, 17-7 is one of the few metals available whose tensile strength increases under a high speed stress; most other metals get weaker. Fig. 6-6 shows Snakes that were tested in hydraulic stretching machines. They deformed, but did not break, and thousands of pounds were required to cause this deformation.

The eye at the bottom of the Snake (Fig. 6-7) is large enough for several harness carabiners, and is formed so that it is not completely closed. This allows the webbing of the rappel seat to be slipped into the eye, if a carabiner is unavailable.

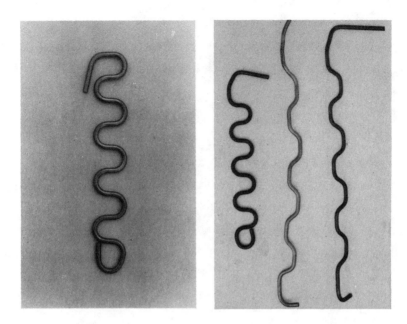

Fig. 6-5   **Fig. 6-6**

Once the Snake is connected to the harness, the rappel rope is attached. This is done by placing the rope into the Snake, and then "winding" it around the bends until it reaches the top (Fig. 6-8). The point where the rope is first inserted is determined by the weight of the rappeller and his equipment, the diameter and length of the rope, the number of ropes, and the rope's condition (i.e., wetness, dirt, or etc.). At the top, the rope is pulled through the "gate" to prevent it from coming loose. This "gate" is quite narrow so that the rope cannot accidentally fall out.

As can be seen, the rope makes contact with the Snake on its horizontal bars. The more bars in contact with the rope, the more friction is produced, and the greater the weight which can be supported. During the rappel, more or fewer bars can be contacted (i.e., the friction in the system can be varied) by winding or unwinding the rope from the lower end of the Snake. This is accomplished by taking the rope over your head or under your feet (Fig. 6-9), which is usually a simple procedure.

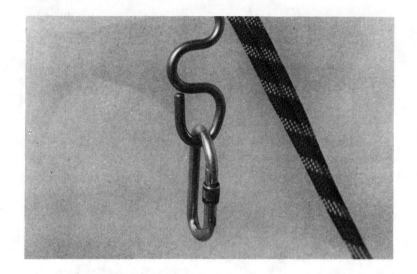

Fig. 6-7

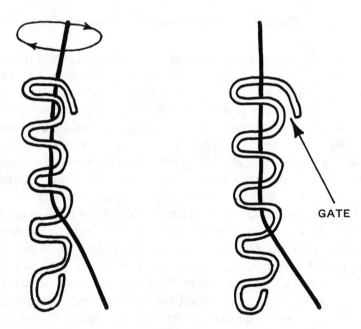

GATE

Fig. 6-8

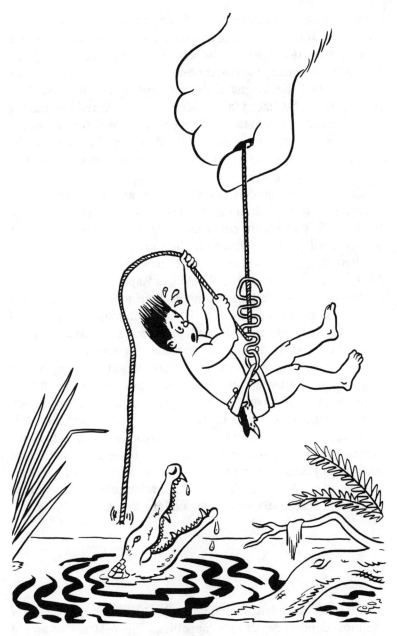

Fig. 6-9

Because the Snake's friction can be varied so simply, it lends itself well to use on moderately long rappels. At the start of a rappel, while the weight of the rope below the rappeller is great, only a few of the top bars on the Snake are used. As the rappeller descends, and the weight of the rope below him becomes less, more bars can be placed in contact with the rope and the braking force required to maintain the rappel speed will remain stable. Just remember that the rope must be lifted over the head in order to adjust friction; so don't use a Snake on a long drop where rope weight is unmanageable.

The ability to add or subtract friction, as needed, provides a simple way to lock the rappel. Just wind the rope around the Snake until all bars are contacted. On double-line rappels this will generally provide enough friction to "stay put." This is not the best method, however. A quicker way is to use the Quick Lock knot (see Chapter 13, "Locking a Rappel").

Besides causing less rope kinking than most descenders, the Snake has one other advantage: The rope running through it does not make sharp bends. This tends to promote increased rope life because the rope fibers are not stretched excessively.

The Snake's main disadvantages are weight and length. Its length makes crawling over certain edges very difficult (see Chapter 11). However, since these rappels occur very seldom, this is of little concern; all rappel devices of any length have this problem. A bigger problem with length is the fact that the descender is nearer your hair. Snakes are more prone than most devices to get caught in the hair, so remember to always carry your prusiks. Also, the Snake does not dissipate heat as well as most large descenders, so go slow on long drops.

FIGURE-8

Figure-8 descenders, sometimes called "8"-Rings, are presently the most popular devices for rappelling. Fig. 6-10 shows an old style which is made of a piece of steel rod formed into a Figure-8

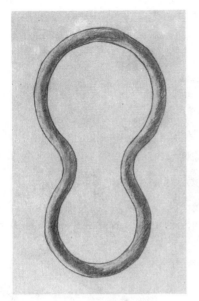 

Fig. 6-10   Fig. 6-11

and welded. This is a very lightweight design which provides good strength. Note that one end of the "8" is larger than the other. This is so that 9mm rope can be used (with the little end up), or 11mm can be used (with the big end up); both with equal efficiency.

Fig. 6-11 shows the type currently available. Depending upon the manufacturer, this type is forged or machined from aluminum alloy and anodized black. Only the larger upper eye is used with the rope; the smaller eye is simply a carabiner hole. Figure-8's are lightweight, but quite strong. They are handy when you have to crawl over edges, because they are short and smooth and don't tend to get hung on things. They are available in different sizes for different size ropes, and on average, weigh 4 oz. (113 g.).

To use a Figure-8, simply insert a loop of rope through the upper eye, bring it under the device and around the backside, and clip your harness carabiner into the bottom eye, as shown in Fig. 6-12. This is a quick and simple procedure which can be done easily in the dark. Connected correctly, the rope cannot accidentally slip out of this descender.

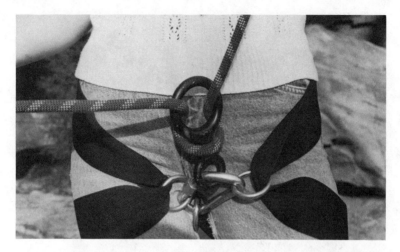

Fig. 6-12

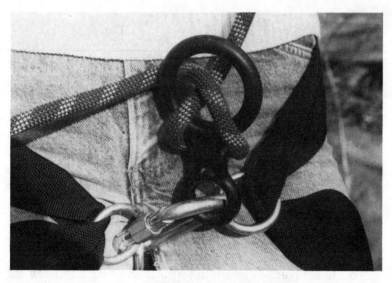

Fig. 6-13

Depending upon the size of Figure-8 and the size of the rope or ropes being used with it, it may or may not provide a safe amount of friction. Most Figure-8's used with single 11mm rope require a lot of braking force – too much for certain rappels. This may necessitate the use of extra friction techniques (see Chapter 13). Rappels with double rope usually provide enough friction, but because Figure-8's cause the rope below them to kink, there can be potential jamming problems when rappelling with a small style.

Figure-8's, because of their low bulk, don't dissipate heat well. It is therefore necessary to rappel slowly on long drops to prevent searing the rope. They do have one great feature however; they provide the rappeller with a simple way to lock the rappel. Just pull the braking end of the rope over and down, between the descender and the standing line, and you will be locked in. You can now safely let go of the rope. To unlock, just reverse the procedure. Fig. 6-13 shows the descender in the locked position. This method of locking requires some muscle, but it does work.

This technique has one inherent danger, however. The rope can slip upward and cinch around the top if the eye (Fig. 6-14). When this happens, you will be stuck – unable to continue down. A very strong person could hold onto the rope above the Figure-8 with one hand, and undo the cinch with the other. People of average strength could not do this! This type of jam can also occur during a normal rappel. It usually happens if the braking end of the rope is held too high, instead of near the hip where it should be. To help reduce the chance of this happening, it might be wise to run the braking end of the rope through the harness carabiner, before it goes to the descender. This also helps to increase the amount of friction in the system. This problem of jamming is a good reason for all rappellers to carry ascending devices, and to know the Shoelace Trick (see Chapter 12).

Figure-8's are not accepted for use by certain fire departments because they must be removed from the rappeller's harness in order to be rigged. However, there is a way to rig without removing the descender from the harness. As can be seen in Fig. 6-15 the rope goes thru the eye of the Figure-8 and is clipped into the harness carabiner. Once in the carabiner, the rope should be looped over the bottom of the Figure-8. This doesn't provide as much friction as a normal rigging, but it does work. It also eliminates the problem of Figure-8 jamming.

Fig. 6-14

Fig. 6-15

Fig. 6-16

Fig. 6-17

Fig. 6-16 shows a SMC Straight-8 descender. It weighs 3.3 oz. (96 g.) and is machined from aluminum plate. The square shape gives it a bit more friction than a normal Figure-8, and also greatly reduces the chance of jamming. The carabiner hole is designed to be used as a Sticht Plate for belaying.

Most Figure-8's will pass single-line knots without too much difficulty. The SMC Straight-8 will pass small knots, but it is sometimes necessary to force them thru by hand. Be careful whenever using small Figure-8's on knotted rope, and rappel with the assumption that there may be a jam. Read the section about "Connecting Ropes" in Chapter 13.

A vastly superior style of Figure-8 is the Figure-8 with Ears. The purpose of the ears, or horns as they are sometimes called, is to prevent the rope from slipping upward and cinching. Eared-8's come in a great variety of shapes and sizes and the larger ones can easily pass most knots. The large sizes have longer life, do less damage to ropes, dissipate heat better, and are easier to lock than standard Figure-8's. When using Eared-8's, be careful when crawling over edges; the ears tend to hang on things.

 Fig. 6-18   **Fig. 6-19**

Figs. 6-17 and 6-18 show a CMI Rescue "8." It weighs 7.8 oz (221 g.) and is made of forged aluminum which has been anodized with a special charcoal colored "hard coat" that is extremely abrasion resistant. It has a two clip-in holes: A large one at the bottom which can be used as a Sticht Plate, and a smaller one which allows rigging without removing the descender from the harness. Fig. 6-19 illustrates how a large carabiner can be placed thru the middle hole, and the rope looped over the bottom tongue of the descender. The long tongue helps keep the rope in position better than an ordinary Figure-8 rigged in a similar fashion, particularly when rappelling double-line. This descender is large enough to pass knots with little difficulty.

Fig. 6-20 shows large and small Russ Anderson Figure-8's with Ears. These descenders are made by SMC, and are machined from ½ inch thick aluminum plate. The large "8" weighs 8 oz. (227 g.), and the small one 6.8 oz. (193 g.). The large "8" also comes in steel, and weighs 24 oz. (680 g.). The larger sizes are designed for use with large diameter rescue ropes, and provide more friction than the small size; even when using smaller sizes of rope. The large sizes will also readily pass knots. The anodizing on the aluminum descenders is extra tough, and resists wear better than most types. Fig. 6-21 shows the small Russ Anderson "8" being used as a Sticht Plate.

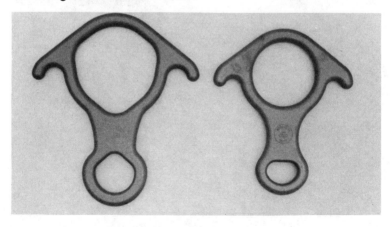

Fig. 6-20

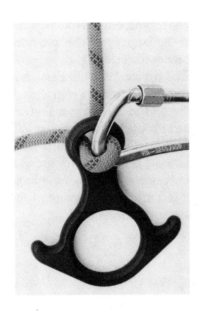

Fig. 6-21

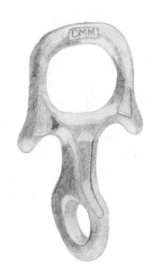

Fig. 6-22

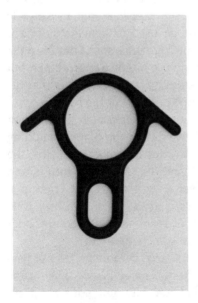

Fig. 6-23

Fig. 6-24

Fig. 6-22 shows a DMM Descender with Ears. Note that the carabiner hole is aligned at 90° to the large eye. This provides better control when using the descender for belaying.

Figs. 6-23 and 6-24 show the CMC Rescue "8" with Ears. It is machined from 1/2 inch 7075 aluminum plate, and given a black "hard coat." It weighs 7.6 oz. (215 g.). The upper eye is 3 inches in diameter and specifically designed for use with ropes over 1/2 in. (13mm), but provides excellent friction when used with smaller ropes. The lower carabiner hole is designed for use as a Sticht Plate, and is presently the only one on any Figure-8 large enough to accommodate 5/8 in. (16mm) rope. Because of the size of this descender, it passes knots very easily.

CARABINER WRAP

This is the official U.S. Army method, and is perhaps the most well known and used of all descending methods. It consists simply of one carabiner, with the rope wrapped around its backside (Fig. 6-25). Enough turns must be used to provide sufficient friction, and the braking end of the rope must spiral out from the bottom of the carabiner. Its best to use a separate carabiner, but the one that holds your rappel harness together can be used as the descender. Just make sure the rope doesn't cut the harness.

A Carabiner Wrap is normally not considered a variable friction descender, but friction can be varied during the rappel by adding or subtracting turns of rope. In order to accomplish this, the carabiner's gate must be opened, and the rope taken thru it, and over the head and under the feet – very much like friction is adjusted in a Snake. Although relatively easy if there is little rope weight below the rappeller, this procedure normally requires both hands, and can be quite difficult in some situations.

Carabiner Wraps, using ordinary sized carabiners, do not pass knots easily, if at all. However, knots can be passed when using very large carabiners – the kind you can put your fist thru. Assume that knots cannot be passed until you prove otherwise.

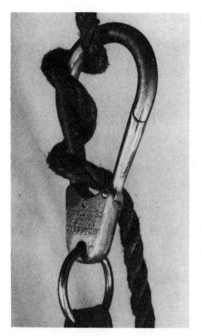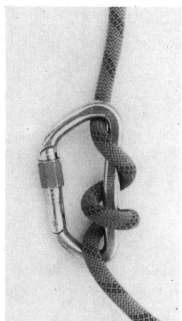

Fig. 6-25

Because of the small volume of metal available to dissipate heat, Carabiner Wraps get very hot very quickly and can easily damage a rope. Aluminum carabiners get rid of heat faster than steel ones, but the difference will hardly be noticed on rappel. Aluminum carabiners are quickly cut in half by dirty ropes, whereas steel ones last much longer. Many steel carabiners are weak, so rappel carefully unless you have one of the better ones.

A method for improving heat dissipation, by using several carabiners in series, is shown in Fig. 6-26. As can be seen, the rope is simply spiraled through the Carabiner Chain; instead of all the turns around a single carabiner. This method is quite safe, so long as the rope is not allowed to run around or against any of the carabiner gates. If an even number of carabiners are used in the chain, and the rope is spiraled in opposite directions through each one, rope twisting and kinking is inhibited (Fig. 6-27). This solves a major problem of Carabiner Wraps.

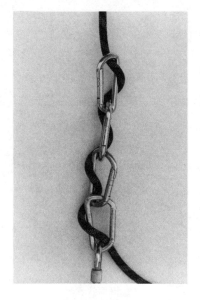

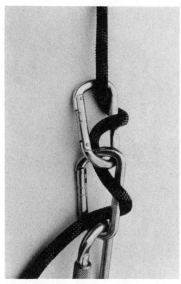

Fig. 6-26   Fig. 6-27

In any Carabiner Wrap system, the gates must point upward (the hinge end goes toward the ground), lest they open up during the rappel and cause an accident. Do believe that gates can open accidently – it has happened many times and caused several deaths. That's why it's wise to use locking carabiners.

BRAKE-BAR AND CARABINER

This descender consists of a brake-bar attached to a carabiner, as shown in Fig. 6-28. It provides a moderate amount of friction, which can be increased by coupling two or more units together with strong chain links (Fig. 6-29).

Although countless rappels have been done with Brake-bar and Carabiner devices, **they are a combination which should be used with extreme caution** – and only with a securing system or belay. Too many people have been killed while using them.

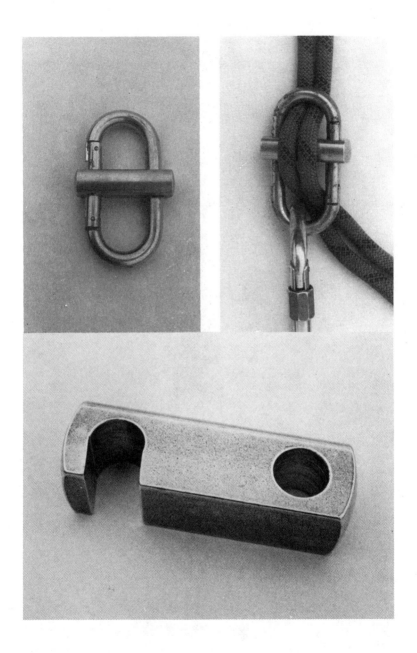

Fig. 6-28

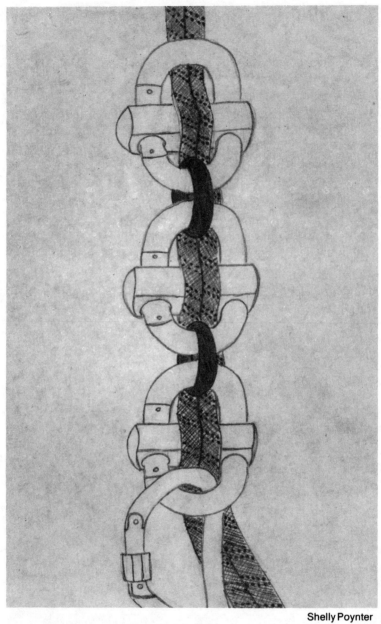

Shelly Poynter

Fig. 6-29

The main problem with this system is breakage of carabiner gate end tabs (Fig. 6-30). They are not very strong, and were never designed to resist large side forces. As shown in Fig. 6-28, when the brake-bar presses against the gate, half the force is held by two end tabs. Depending upon the brand of carabiner, end tab strength can range from 500 to 2,500 lbs. (227 to 1,134 kg.). This may sound like a lot, but it doesn't take into account flaws in some carabiners, or metal fatigue. What it all comes down to is this: During a hard rappel the tabs can snap off and let the rappeller fall. It has happened numerous times!

The use of two or more units in series, as shown in Fig. 6-29, can increase system friction and safety. If one gate should break, there are others to hold you – maybe. Of course, as the rope slides past the broken gate, it could get cut. Perhaps the best way to insure maximum safety is to make sure all gates point upward. The hinge end must point down. This orientation locks the gates closed and maintains maximum carabiner strength.

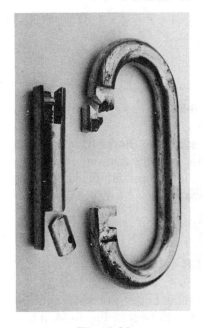

Fig. 6-30

Fig. 6-31

The next biggest safety hazard with brake-bar devices is connecting into them on the backside of the brake-bar, as shown in Fig. 6-31. Sometimes this mistake will support enough weight to get you over the edge, and then the brake-bar will flip open. This happens most often when using slant slot brake-bars (Fig. 6-32); they can grip tightly when hooked up incorrectly. Brake-bars with square cut slots (Fig. 6-33) are much less apt to fool the rappeller, and are the most desirable type to use.

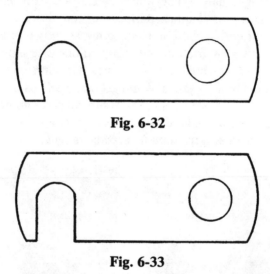

Fig. 6-32

Fig. 6-33

Once again, it must be stressed that Brake-bar/Carabiner descenders should be used with extreme caution! Carabiner Wraps are much safer; their only important disadvantages being rope kinking and poor heat dissipation.

CARABINER CROSS

This system was developed by mountaineers to avoid carrying the extra weight of brake-bars. As can be seen Fig. 6-34, it uses nothing but carabiners.

The body of this device must always consist of two carabiners with the gates on opposite sides – this is of extreme importance! Both gates should also point upward. The "brake bar" can be just one carabiner, but two are safer and provide much more friction. The gate of the braking carabiner is always placed on the back side of this rig – the rope must never run across it. For more friction, two or more Carabiner Cross units can be coupled together by carabiners to form a chain of descenders.

Because the gates on the carabiners are either unstressed, or backed by solid bars, this system is substantially safer than Brake-bar and Carabiner descenders. However, it is a complicated system to rig, particularly in the dark, and it dissipates heat no better than Brake-bar systems. It has some advantage over Carabiner Wraps, it doesn't kink ropes as badly; but in light of the rigging complications, it is a poor choice of descender. Carabiner Wraps should always be considered first. In most rappel situations, simplicity means safety!

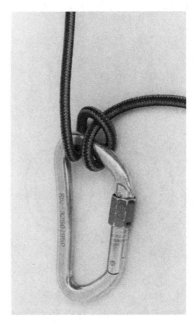

Fig. 6-34   Fig. 6-35

HALF RING BEND

Also known as the Friction Hitch, Munter Hitch, or Italian Hitch, this is another way to rappel using only one carabiner (Fig. 6-35). As can be seen in Fig. 6-36, it is very simple to rig, but it does provide great amounts of friction – far more than a Carabiner Wrap. It works so well that it is the official U.I.A.A. method for providing friction for belaying.

Because the rope rubs against itself, the Half Ring Bend is not a desirable method for very long rappels, or repetitive rappels on the same piece of rope. However, it is a very safe way to rappel under some circumstances. This rig obviously does not dissipate heat well, so remember to go slow.

This is one case where turning the carabiner gate down is okay, and perhaps advisable. Also, it doesn't matter which end of the rope goes to the anchor or braking hand, the Bend will reverse itself on demand and provide friction in either direction of rope pull. This is one of the reasons it makes a good belay device.

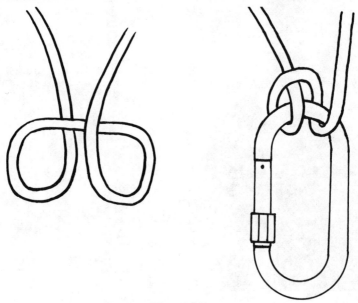

Fig. 6-36

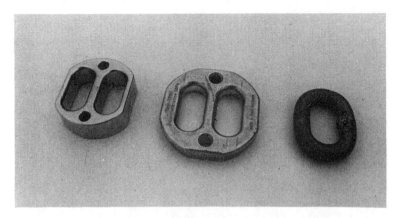

Fig. 6-37

STICHT PLATE

This is a most ingenious device which is used primarily for belaying, but can also be used for limited rappelling. Styles of Sticht Plates vary from simple chain-links for use on single line, to large dual-slot units for double rope technique (Fig.6-37). They are usually made from cast or machined aluminum, and weigh an average of 1.4 oz. (40 g.)

Fig. 6-38 shows a Sticht being used for rappelling. A loop of rope is inserted through the slot in the Plate, and a carabiner is clipped into the loop. This carabiner is then connected into the seat harness. Rappelling is accomplished by merely pulling on the braking end of the rope; as with other descenders. A number of Figure-8 descenders have carabiner holes which double as Sticht Plates. Fig. 6-21 is one example. This arrangement is normally used for belaying, not rappelling. A Sticht belay setup is shown in the cover photo of Chapter 10.

The slots in Sticht Plates must be the proper size for the rope being used; otherwise too little or too much friction will be produced. Most plates are stamped with the rope size. Slots designed for 7/16 inch (11mm) rope are about 1.25 inch (32mm) long, and .58 inch (15 mm) wide. Dual-slot plates can, of course, be used for single-line work.

Fig. 6-38

Because most Sticht Plates contain very little metal, they dissipate heat poorly. This means that you should avoid long, fast rappels. Another danger is using the Plate as shown in Fig. 6-39. In this arrangement the carabiner is used only as a brake-bar; the rappeller connects his harness to the Plate with a piece of small diameter cord. This is insane!

Fig. 6-39

PRESSURE PLATE

A drawing of a rudimentary Pressure Plate is shown in Fig. 6-40. It consists of two metal plates which are hinged on one side, have handles on the other side, and rope guides in the middle. To use this device, it is clamped around the rope, and the handles are squeezed in order to obtain friction for braking.

When extremely long rappels are made, using descenders like Racks or Spools, much frictional heat is generated, and rappels are done slowly in order to prevent melting the rope. In order to descend rapidly, descenders which do not apply pressure to the rope, until needed, must be used. Pressure Plates may lightly contact the rope during swift descent, but do not actually get hot until braking is applied by squeezing the handles.

At the present time, Pressure Plates are still experimental. The primary problem is that most of them dispel heat poorly, and when squeezed tightly during braking, actually crush fibers and bring about weakening of the rope. The design illustrated has poor ventilation, permitting heat in the rope to build up rapidly. Because the rope is clamped tightly on either side, plastic deformation of the fibers results; as does excess heat from interfiber friction. The heat buildup, plus the flattening of fibers, can easily cause a rope to melt and break.

To improve heat dissipation and reduce rope-crushing effect, the interior of the plates can be equipped with brake-bar style projections, in an offset arrangement. Fig. 6-41 shows a side view of such a design, with the "jaws" open. With this design, the rope is not crushed, and heat dissipation is vastly improved. There are other problems, however.

A secondary problem encountered when using Pressure Plates is that of providing bottom belay. With most descenders, pulling on the braking end of the rope, below the rappeller, will bring the rappel to a halt. Because Pressure Plates generate friction only when their handles are squeezed, bottom belays are useless – and can be dangerous if used with the style Plate shown in Fig. 6-41. Because Pressure Plates are used for making very long, fast rappels, top belays are also usually impractical.

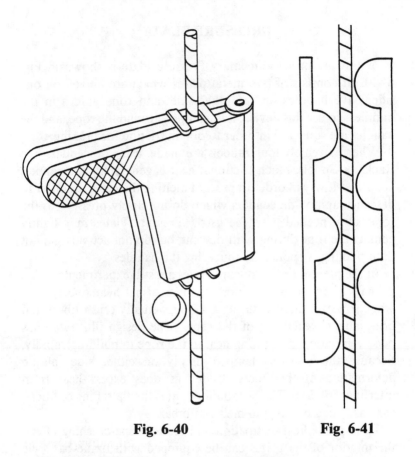

Fig. 6-40   **Fig. 6-41**

Because of their present stage of development, and proven danger, Pressure Plate rappels should only be undertaken by individuals well versed in rappelling and engineering.

DESCENT-MASTER

Manufactured by the Atlas Safety Equipment Company, the Descent Master is made of cast aluminum, weighs 2 lbs. (907 g.), and features an integral "dead-man system." It will stop a rappel instantly if the rappeller loses control.

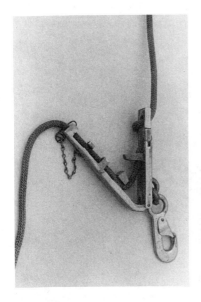

Fig. 6-42   Fig. 6-43

As shown in Fig. 6-42, the Descent-Master has a control handle, integral snaphook for connecting to the rappel harness, and a carabiner type braking plate. The rope is inserted through the gate on the braking plate, and wrapped one or two times around the braking shaft. One turn will allow loads of 70-349 lbs. (32-158 kg.) to be lowered; while two turns will support 350-400 lbs. (159-181 kg.). Instructions for use are printed on the unit.

To descend, the control handle is pushed toward the ground slowly, until the desired descent speed is reached (Fig. 6-43). Rate of descent is determined primarily by the weight of the rappeller, and the angle of the control handle. Control is possible because of the friction variation which occurs as the rope changes angle around the center rope guides.

To slow the rappel, merely move the handle upward, toward the braking plate. This occurs automatically if the hand is removed from the handle; thereby providing instant "dead-man" locking of the rappel (Fig. 6-44). This automatic upward movement of the handle occurs because of the friction between the rope and the

Fig. 6-44

rope guides, and the angle the rope makes between the center guides. With the handle up, the Descent-Master will hold a load of 350 pounds with one turn of rope around the braking shaft; and 600 lbs. (272 kg.) with two turns. At weights greater than these, the unit will slide down the rope in a dynamic belay fashion.

Early model Descent-Masters could be controlled from the ground. By pulling on the rope below the unit, the control handle could be pulled down and the load would descend. Pulling even harder would stop this movement because of friction increase around the braking shaft. These early models also featured kernmantel rope with a steel core. Because of various problems these two features are not incorporated in present models.

Descent-Masters are equipped with a kernmantel rope, the diameter of which is critical to the function of the device. If too large or stiff a rope is used, the descender may jam. If the rope is too small, it may slide at low loads. The diameter of the rope used is approximately 7/16 (11mm), and it comes with a snaphook attached to the end.

The rope is held on the Descent-Master by several slotted rope guides which face in opposite directions. It is secured onto the control handle by a lock pin through the first guide. On the braking plate, the rope runs through a single guide and is secured by the breaking shaft. Getting the rope into and out of the guides is not easy, and requires many seconds. If speed is a factor, such as in emergency use, the Descent-Master should be attached to the rope before the need arises.

Because of the bulk of the braking arm, the Descent-Master dissipates heat very well. However, because the rope spirals around the braking shaft, the rope below the unit kinks badly. This, combined with the general roughness of the metal surfaces, can result in slow, sluggish rappels at low loads.

The snaphooks at the ends of the Descent-Master ropes are secured to loops formed by the Thimble/Swage Sleeve method. These should be checked often for rope slippage. For more on this matter, see the "Sky Genie" section. Because Descent-Masters have been known to break during a rappel, they should be used with extreme caution.

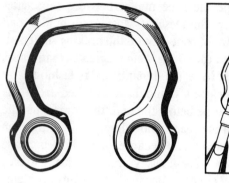

Fig. 6-45

Fig. 6-46

FROG

The Edelrid Frog (Fig. 6-45) is a belay device which can be used for rappelling. It weighs 2.3 oz. (65 g.) and is made of aluminum which is anodized in various colors. It is designed to be used with seat harnesses which require two carabiners, although it can be used with regular harnesses. It is normally used with the Half Ring Bend (Fig. 6-46), although it will work if rigged like a Carabiner Wrap. Because of the Frog's low bulk, and the rope rubbing across itself (when rigged with the Half Ring Bend), it is best used on short, slow rappels.

SPOOL

Two types of Spools are shown in Figs. 6-47 and 6-48. They are nothing more than large spools made of metal tubing, with end plates and attachment arms welded on. Some, such as the Corkscrew Spool (Fig. 6-48), have rope guides; but these are not absolutely necessary. Small spools, which are vertically oriented, are sometimes called Spindles. To prevent the rope from coming off, they always have rope guides. Spools are normally made of steel, but may be made of aluminum if weight is critical.

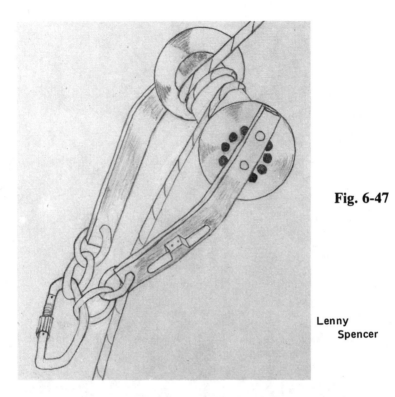

Fig. 6-47

Lenny
Spencer

Because of size, most Spools dissipate heat extremely well, making them useful for long, fast rappels. They are very safe and probably do less damage to the rope than any other type descender.

An important feature of most spools is the ability to pass knots. Styles without rope guides will slide over the largest knots with ease. Corkscrew types have to have the rope guides removed from around the rope before knots can be passed. Some spools, with small rigid guides, can't handle large knots. Spools are also valuable because they can handle a wider range of rope sizes than any other descender.

The friction of Corkscrew Spools, and some other types, can be easily varied by removing the braking end of the rope from its guide, and adding or subtracting turns from around the drum. This trick can't be used on types like Fig. 6-47, which have two attachment arms running to the harness.

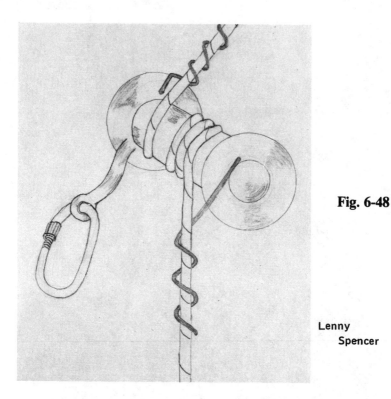

Fig. 6-48

Lenny
Spencer

The main disadvantage of Spools is their bulk and weight. They also cause rope kinking because the rope is spiraled about them. However, these disadvantages are usually offset by their excellent performance on long rappels. Also, because of the Spool's large size, there is plenty of space for things like cleats, which are a big help in locking a rappel.

A unique variation on the Spool concept is the Rope Rider, made by Forrest Mountaineering. The regular size, designed for standard rappel ropes, is shown in Fig. 6-49. It weighs 14.5 oz. (411 g.). The Junior Rope Rider, shown in Fig. 6-50, weights 10.5 oz. (298 g.). Fig. 6-51 shows a Rope Rider in use.

The drum of a Rope Rider is hollow, for good heat dissipation, and is made of 6061-T6 aluminum. The connecting arm is steel and is screwed into the drum and secured by a lock nut and lock pin. The lower rope guide is simply a pin projecting from the drum.

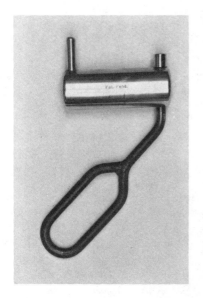

Fig. 6-49   **Fig. 6-50**

Rope Riders are simple and fast to rig. The rope is given three turns around the drum, and the braking end is run thru the harness carabiner. Friction can be easily adjusted, while on rappel, by merely winding or unwinding the rope from around the lower portion of the drum. Like other spiraling type descenders, Rope Riders tend to cause the rope below them to kink. However, because of the diameter of the drum, rappels are usually very smooth, with little problem of jams.

The Junior Rope Rider is designed for use with 5/32 inch (4mm) diameter Kevlar Rope, which is about the size of parachute cord. The rated strength of this rope is 2,600 lbs. (1,179 kg.). This type Kevlar rope should never have knots tied in it, nor be bent around anything less than ten times its own diameter, but it can be used with the Rope Rider because of the diameter of the drum. The purpose of using Kevlar rope with the Junior Rope Rider is so that firemen can carry a complete self-rescue system at all times. The Rider, plus several hundred feet of rope, can easily be carried in a pants pocket.

Fig. 6-51

Although Kevlar is extremely strong, it is possible for it to be broken by quick rappel stops because of lack of elasticity. Anchors can be broken or pulled out for the same reason. When using this type rope, rappel as smoothly as possible. During practice, always use a belay. The Kevlar rope provided with the Junior Rope Rider should not be run over sharp edges, so always use heavy protectors on the rope, and pad sharp edges.

Kevlar will withstand high temperatures, and will not melt from a fast rappel. It can even take direct flame for a few seconds; but don't count on being able to rappel thru hot flames without damage to the rope. Long and fast rappels will get the descender extremely hot, so be careful about touching it with bare hands. Also, bear in mind that such small diameter rope may be hard to grip during rapid braking. Never use this system without wearing very heavy leather gloves. Always put enough turns around the drum so that braking can be achieved without burning thru the gloves. This system was designed as a "last resort" backup; not a desirable way to rappel. Use it very carefully!

BANKL PLATE

The Edelrid Bankl Plate (Fig. 6-52) is something of a cross between a Sticht Plate and a Figure-8. It is designed for use with the type harness which has two attachment points. However, it can be used with standard harnesses by clipping both carabiners onto the same harness loop. The main virtue of the Bankl Plate is that it keeps the rappel ropes separated as they slide through. It also inhibits rope twisting and kinking. The Bankl Plate can be used with one rope, as shown in Fig. 6-53.

The Bankl weighs 3 oz. (88 g.) and is made of aluminum; but because of its small size, should be used slowly on drops of any length. Directions for use are stamped on the side.

 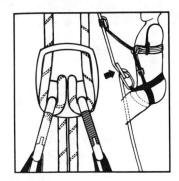

Fig. 6-52

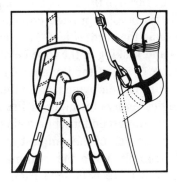

Fig. 6-53

 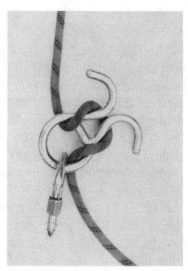

Fig. 6-54   **Fig. 6-55**

LONGHORN

Made by Mountain Safety Research (MSR), this aluminum descender weighs 5.2 oz. (146 g.), and is shown in Fig. 6-54. It operates somewhat like a Figure-8, only much better; it won't jam like a standard Figure-8! As can be seen in Fig. 6-55, the rope is simply pushed through the eye of the Longhorn and looped over its horns. If this doesn't provide enough friction, another loop is wrapped around the horns (Fig. 6-56). This extra loop also eliminates twisting and kinking of the rope below the rappeller.

Because of the ease of use of the Longhorn, and the fact that it can be hooked up without removing it from the harness, it is one of the more useful descenders for night or fire rescues. There is no danger of losing your descender while trying to rig for rappel. Longhorns also provide one of the simplest and safest means of locking a rappel: Simply wrap the braking end of the rope around the horns until there is enough friction to hold you in position (Fig. 6-57).

Fig. 6-56   Fig. 6-57

The eye of the Longhorn is fairly large; thus can pass small knots and rope kinks. On long rappels it is best to go slow; Longhorns don't dissipate heat well because of their light weight and silver color. They are about on par with Figure-8's in this respect. Be careful when crawling over edges; the long horns tend to hook on things. During a rappel, there is no danger of the rope coming off the horns; but this could happen right before the rappel, if one was careless. Before backing over the edge, make sure the Longhorn is rigged properly.

PIERRE ALLAIN

The Pierre Allain (Fig. 6-58) is one of the oldest descenders. The hole in the handle is for the harness carabiner. The rope is brought over the rear horn, and curled twice around the body. Braking is accomplished by pulling on the braking end of the rope. Because its very easy for the rope to come off the Allain, it must be used with extreme caution. It is best left in the museum!

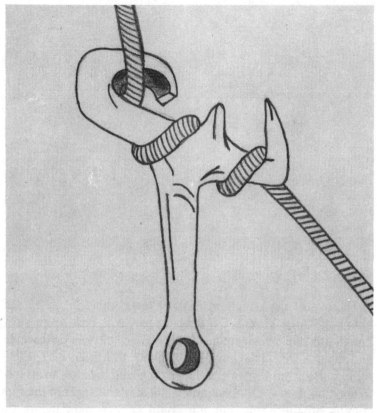

Fig. 6-58

Shelly Poynter

ROBOT

Made by Bonaiti, this descender weighs 5 oz. (140 g.) and consists of a hollow aluminum brake-bar, which is permanently attached (on one side) to a small aluminum rack (Figs. 6-59 and 6-60). The hole at the bottom of the rack is for the harness carabiner, and the horns are used for gaining extra friction. A Robot can be used with 5mm-13mm rope, and can also be used as an ascender, when rigged properly.

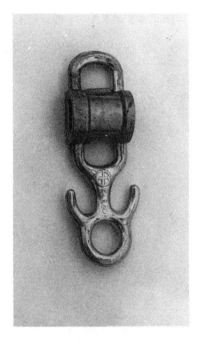 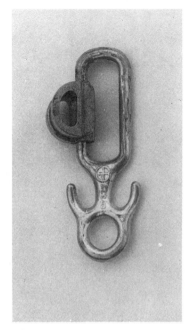

Fig. 6-59   Fig. 6-60

Because of its small size and silver color, this descender doesn't dissipate heat well, but it is more efficient than devices of similar weight because of the hollow brake-bar. It should be rappelled on carefully – and the neck and frame checked often for cracks.

Fig. 6-61 shows the Robot rigged for a normal rappel. A Prusik is shown attached above the Robot for securing the rappel and it works best if the Prusik sling is run through the hollow brake-bar. When the Prusik locks, it pulls the brake-bar upward, putting great friction on the rope by means of a tooth on the upper side of the bar. The Prusik sling should be tied with a Slipped Sheet Bend to expedite release.

Fig. 6-62 shows the Robot rigged for double-line rappelling. If extra friction is needed, the ropes can be run across the horns at the lower end of the rack. If extra friction is needed when rappelling on small ropes or single-line, the descender can be rigged as shown in Figs. 6-63, 6-64, and 6-65.

Fig. 6-61

Fig. 6-62

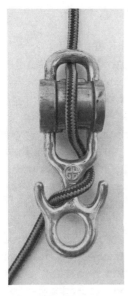

Fig. 6-63

Fig. 6-64

Fig. 6-65

PETZL "CLASSIC"

Made by Petzl, this descender is used for rappelling on single rope of up to 12mm diameter. It weighs 9 oz. (260 g.). Fig. 6-66 shows one of the older models in use. The rope runs across two spools (capstans) which distribute the friction over the mantel, thereby reducing mantel slippage. This descender does not cause excessive twisting or kinking of the rope. The spools on the Classic are made of aluminum, and are prevented from rotating by slots on their backsides (Fig. 6-67).

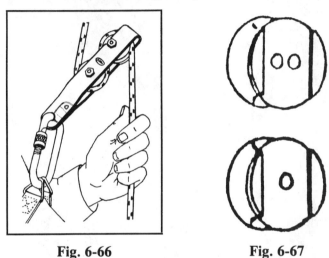

Fig. 6-66   **Fig. 6-67**

Fig. 6-68A shows the newest model Classic. Its frame is shown rotated open, so that the rope can be inserted. The frame is made of anodized aluminum, and the screws, pins, and spring are made of stainless steel. A "rapid clip" type gate, on one of the frame halves, allows the frame to be opened without removing the unit from the harness. Instructions for use are stamped on the frame. Fig. 6-68B shows a side view of the unit.

Fig. 6-68C shows a way to rig when less friction is needed. This is especially useful when rappelling on large ropes. If more friction than normal is needed, like when rappelling on small diameter ropes, the rigging shown in Fig. 6-68D can be used.

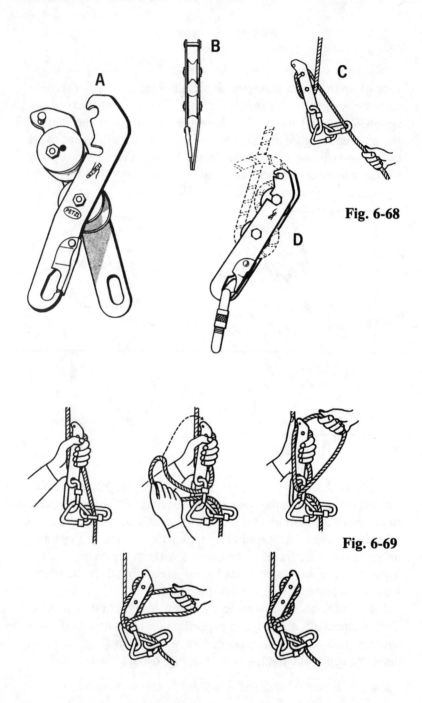

Fig. 6-68

Fig. 6-69

By pulling a loop of rope through the harness carabiner, and looping it over the top of the descender, the rappel can be easily locked (Fig. 6-69). For this method to work properly, it is necessary to run the braking end of the rope through a carabiner, as shown in the drawings.

Petzl descenders are designed for caving, but are very useful in any type of single-line rappelling. Many rappels can be done on them, because when the spools become excessively worn on one side, they can be reversed in order to provide a fresh surface. Disassembly is easily done with two wrenches, and spare parts are readily available from the factory.

STOP

This descender is also made by Petzl, and like the Classic unit described previously, is designed for single-line rappels on ropes up to 12mm diameter. It weighs 11 oz. (315 g.). Fig. 6-70 shows an old style STOP in use, while Fig. 6-72 illustrates the new type. The STOP has a "dead-man" lever which must be squeezed in order to rappel. If the rappeller loses control, and removes his hand from the lever, the descent is automatically halted (Fig. 6-71). With a hand on the lever, this descender functions like most others, and braking is accomplished by pulling down on the braking end of the rope. Although braking can be done by only squeezing the lever, and not pulling on the braking line, this is not the recommended mode of use. It could lead to loss of control, and the rapid stops could cause damage to the rope or anchor.

Braking is effected by means of a stainless steel tooth on the lower spool, which is attached to the end of the lever. When there is no hand on the lever, rope friction around the lower spool rotates it a few degrees, and squeezes the rope between the tooth and the upper spool. Restraint is almost instantaneous on normal sized ropes, and there is substantial friction on ropes as small as 1/4 in. (6mm); although with such small diameters some force may have to be applied to the braking end of the rope.

 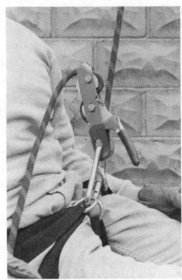

Fig. 6-70   **Fig. 6-71**

STOP frames are made of anodized aluminum, and the spools are bare aluminum. Screws and pins are made of stainless. The STOP is rigged in the same manner as the Petzl Classic, by rotating the frame open and inserting the rope. Like the Classic, this can be done without removing the unit from the harness, thanks to the "rapid clip" on the carabiner hole. Instructions for use are stamped on the frame.

On certain rappels, where only one hand is free for braking, it may be desirable to disable the "dead man" system. This can be done by clipping a carabiner into the hole on the rear of the lever arm (Fig. 6-72C), forcing the unit to function as a normal descender. If more or less friction than normal is required, the STOP can be rigged as shown in Figs. 6-72A and B.

Like the Petzl Classic, the components of the STOP are easily replaced when worn. Because of this "repairability," ease of use with different sized ropes, and light weight, the STOP is perhaps the most desirable descender available when automatic braking capability is required.

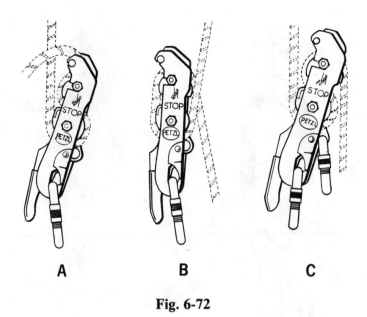

A   B   C

Fig. 6-72

Be very careful, however, when descending rapidly and stopping suddenly with the "dead man" lever. The tooth can actually take a "bite" out of soft, braided type ropes. This descender should be used only on kernmantel rappel ropes, and the braking lever should never be used for making sudden stops – except in emergencies.

SEILBREMSE

Made by Bachli, this descender weighs 3 oz. (85 g.) and is forged of steel, which is then blued. It can be used with ropes of 5mm to 13mm diameter. Fig. 6-73 shows a Seilbremse rigged for normal rappelling. Rigged in this manner, it also does quite well as a belay device. Fig. 6-74 illustrates how to get maximum friction when rappelling single-line. The rope is run across itself, and around the harness carabiner. Although this method provides much friction, it is not desirable because of the rope surfaces rubbing together.

Fig. 6-73

Fig. 6-74

Fig. 6-75

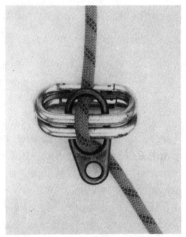

Fig. 6-76

Fig. 6-75 shows how to rig the Seilbremse when rappelling on large ropes and double-lines. A carabiner is placed across the device to act as a brake-bar. This method is similar to the Carabiner Cross illustrated in a previous section. Make sure that the carabiner gate is on the backside of the Seilbremse – under no circumstance let the rope run across the gate! Fig. 6-76 shows using two carabiners for extra friction. On double-line rappels with 11mm rope, this rig provides adequate friction.

Because of the small size of the Seilbremse, and the fact that it is made of steel, it does not dissipate heat well. It should be used carefully on long rappels. The main virtue of this device is its long life. It will outlast several aluminum descenders. It should be used only with full-size carabiners; preferably steel ones. Do not use lightweight aluminum carabiners. One long rappel on a muddy rope could cut them severely.

RUAPEHU

Made by Ruapehu Mountain Equipment Co., this descender is one of the most versatile available. It is made of aluminum and is clear anodized, and weighs 5.4 oz. (154 g.). Because of the many different ways it can be rigged, great amounts of friction are available with most rappel ropes. Fig. 6-77 shows a Ruapehu.

Figs. 6-78, 6-79, 6-80, and 6-81 illustrate a few of the ways to rig for single-line rappelling. Figs. 6-82 and 6-83 show how the descender can be used for double-line rappelling. Rope twisting and kinking does not occur when the Ruapehu is rigged as shown in 6-79, 6-80, 6-81, and 6-83. Note that Figs. 6-80, 6-81, and 6-83 show the Ruapehu being used as a Sticht Plate.

The horns on the Ruapehu can be used for locking a rappel. Simply wrap the rope around the horns a couple of times as shown in Fig. 6-84. The rappel can also be locked by pulling a loop of rope through the harness carabiner and hanging it over the horns (Fig. 6-85). This latter technique is useful if the horns already have two lines around them.

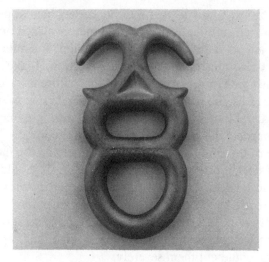

Fig. 6-77

The metal in the Ruapehu is of thick cross section. Ropes are not damaged as much by this device as with some of thinner material. Ruapehus are very strong and safe, and can be rigged without removal from the harness. They are especially useful when doing both belaying and rappelling on double-line.

Fig. 6-78

Fig. 6-79

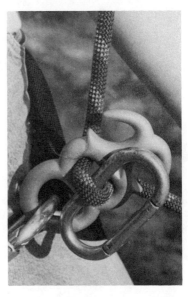

Fig. 6-80   Fig. 6-81

Fig. 6-82

Fig. 6-83   **Fig. 6-84**

Fig. 6-85

Fig. 6-86

SKY GENIE

The Sky Genie is manufactured be Descent Control, Inc., and comes in three sizes. The largest, shown in Figs. 6-86 and 6-87, weighs 1 lb. (454 g.). It uses 1/2 inch (13mm) rope, while the smaller units use 3/8 inch (10mm) and 5/16 inch (8mm) rope. The ropes supplied with the units are white braided nylon and have attachment loops formed at each end. The 1/2 inch and 3/8 inch ropes also come with heavy rope protectors.

Sky Genies are made of aluminum, and are a two-piece device consisting of a sturdy center spindle (Fig. 6-88) and a tubular cover (Fig. 6-89). The cover serves primarily to hold the rope on the spindle; but also helps keep foreign objects out of the descender. The covers for the two larger units have a gap down one side which allows the Sky Genie to be attached at any point on a rope. The cover is placed over the rope and slipped up over the spindle, and held in place by a small thumb nut. This cover must be securely in place before rappelling!

Fig. 6-87

As can be seen in Fig. 6-88, the rope is wound about the spindle and placed into slots at either end. There are two slots at each end, to facilitate double-line rappels. Normally, the rope is given at least two turns around the spindle, but more can be added if needed. They cannot be added, or subtracted, once you are on rappel. Basic instructions for rigging are printed on the side of the cover. Rappels can be locked by looping the braking end of the rope over the top of the Sky Genie, between the standing rope and the upper eye (Fig. 6-90).

Although Sky Genies are heavy, they don't dissipate heat well because of the cover restricting ventilation. Therefore, it pays to go slow. In use, the side gap should not be covered with the hand, as this would restrict ventilation even more.

When used properly, the Sky Genie works satisfactorily. Its prime problem is one of the rope slipping out of an upper slot and making a half turn around the top of the spindle. This only occurs in very hard, fast stops, with heavy loads. Generally, the worst effect of this is a bit less friction. Under extreme circumstances the cover can fly off, letting the spindle come off the rope. This should not happen in normal rappels, however. Because kinks or knots in the rope might jam the descender, or possibly dislodge the cover, rappels should be done slowly and carefully. It might also be wise to place a securing device on the rope, above the descender. For more information on safety systems, see Chapter 13, "Securing a Rappel."

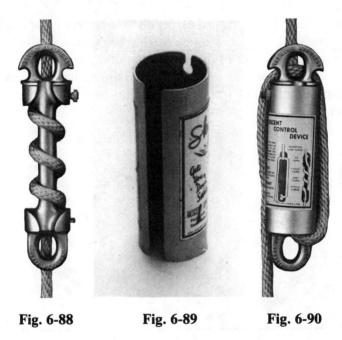

Fig. 6-88   **Fig. 6-89**   **Fig. 6-90**

Because the Sky Genie requires its cover to be operative, if the cover is lost, the spindle is useless. Be extremely careful about losing it over the edge when hooking up. If this should happen, remember that you can use your harness carabiner as a descender (see "Carabiner Wrap").

The attachment eyes at the end of the Sky Genie rope must be checked often for slippage of the line through the swage sleeves. Such inspection is advisable with any ropes which have end loops formed by the Thimble/Swage Sleeve method. This method works well with manila rope, but with nylon – because of its stretch – there is a remote chance that the rope could be pulled through the sleeves during heavy loading. In certain cases, it might be wise to replace these eyes with Figure-8 loops.

RACK

The brake-bars maligned earlier do have a place in rappelling, and that is on a Rack. This device has been made in all shapes and sizes, and is one of the most useful descenders. Most racks are designed to hold four to six brake-bars, but in order to dissipate the heat of very lengthy rappels, long racks have been built which contain many bars.

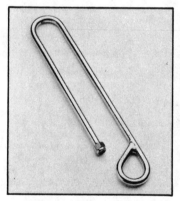 

Fig. 6-91

Fig. 6-91 shows a SMC Rack, which consists of a heavy 304 stainless rod frame and lots of brake-bars. The large eye at the bottom of the frame is for carabiner attachment, and it is welded closed for extra strength. This rack is available in 5 or 6-bar models, and is designed to use standard size brake-bars. SMC makes bars of solid aluminum, and of tubular steel or stainless.

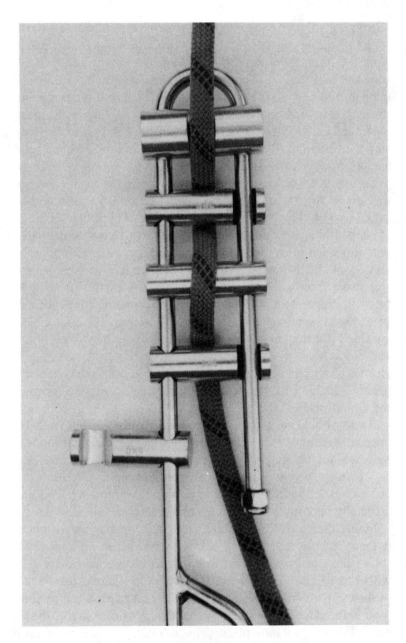

Fig. 6-92

Rigging a Rack, as seen in Fig. 6-92, is very simple. To start with, all the brake-bars, except for the top one, are unclipped from the frame. The rope is placed against the top bar, passed thru the rack frame, and then the second bar from the top is clipped into place. The rope is brought back thru the frame in the opposite direction, and bar #3 clipped in. The process is continued until the rope is in contact with enough bars to provide sufficient friction. When rigged correctly, the rope does not rub against the top of the Rack frame.

Because the bars face in opposite directions, the rope is forced to zig-zag thru them. Be sure that the rope only contacts the brake-bars on their front side – not the slotted backside. Letting the rope run against the backside can be lethal because the bars will flip open under load (see the "Brake-bar and Carabiner" section of this chapter). To reduce the chance of bars being rigged incorrectly, the second and third bars from the top should be of the square cut slot variety. For maximum safety, all the bars should be square slotted; although it is normally safe to use a slanted slot top bar – if it fits tightly on the frame. Figs. 6-91 and 6-92 show a slant slot SMC top bar with a "training" groove. This bar is made larger in diameter than the others in order to dissipate heat better. The groove helps keep the rope in the middle of the Rack.

Racks should be always be rigged with more friction than deemed necessary. In situations which demand much friction, such as short rappels, small diameter ropes, carrying a backpack, etc., it might be wise to engage all the bars, since it is easier to decrease friction than increase it while on rappel. This is especially important until skill in using a Rack is gained.

Manipulating the brake-bars, while on rappel, is very simple if there is little rope weight below the rappeller. Fig. 6-93A shows the side view of a 6-bar Rack, with the rope in contact with three bars. To increase the available friction, bar #3 is pushed upward with one hand while the other hand moves the rope to the side (Fig. 6-93B). When enough clearance is gained, bar #4 is flipped into place (Fig. 6-93C). To decrease the friction in the system, the procedure is reversed.

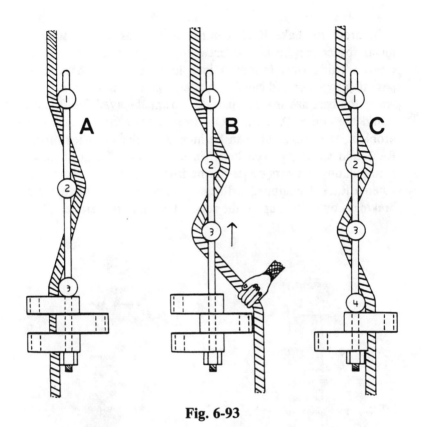

Fig. 6-93

Extra friction can also be obtained, during the rappel, by manually pushing the bars upward with the balance hand. This forces them closer together, causing the rope to bend around the bars at a greater angle, thus increasing the amount of rope/metal contact. Control can be "fine tuned" in this manner. A similar effect can be accomplished by simply pulling the braking end of the rope upward, which also forces the bars closer together. If the rope weight below the rappeller is not excessive, this is perhaps the best way to do rapid braking with a Rack.

When braking by pulling upward on the rope, bear in mind that a standard bottom belay is contraindicated. If the rappeller is trying to pull up on the rope, and the belayer is pulling down, an accident could result.

In order to make Rack control more consistent, a tubular metal spacer can be placed between the two upper bars (Fig. 6-94). This spacer (about ⅝ inch long) keeps the bars from jamming together and causing sudden increases in friction. The wider the bars are spaced apart, the lower the available friction – and the cooler they run. On drops where rope weight is a problem, the brake-bars can be hinged on the short leg of the Rack and spacers placed between every bar. This can make manipulating the bars easier during friction changes. However, when a Rack is equipped with spacers between every bar, rapid braking – by pulling up on the rope – becomes impossible.

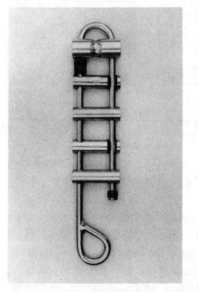 

Fig. 6-94	**Fig. 6-95**

Brake-bars are attached to a Rack by removing the lock nut on the end of the frame and slipping the bars on. Normally, they are taken over the top of the frame and placed on the same side as the carabiner eye. On this long side they are less in the way when not being used, thus permitting the upper bars to be spread further apart when a reduction in friction is needed.

A most important reason for hinging brake-bars on the long side of a Rack is shown in Fig. 6-95. When friction is needed during an emergency, the braking end of the rope can be wrapped around the lower end of the Rack frame. This provides much friction and works well – if the rope weight is not excessive.

Racks can be locked by looping the rope over the top, between the frame and standing line, in a manner similar to that shown in Fig. 6-90. Using several turns, or tying a loop of the braking line around the standing line, will increase the security.

Most Racks are constructed similarly to the one shown in Fig. 6-91. Other designs vary greatly from this concept. One such type is shown in Fig. 6-96. It is designed for military use, and is formed from aluminum sheet and tubing. As can be seen, the "brake-bars" on this Rack are fixed, and do not pivot. The rope has to be threaded thru them before rappel. Because this Rack provides no way to vary friction during the rappel, except by pulling on the braking end of the rope, it must be rigged with the proper amount of friction – before going over.

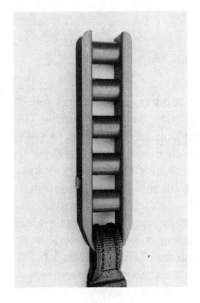

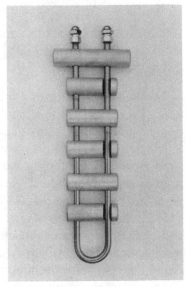

Fig. 6-96   **Fig. 6-97**

Another unique design is the RSI Rescue Rack, by Rescue Systems, Inc. (Fig. 6-97). The brake-bars are solid aluminum, the frame is stainless, and the total weight is 2.1 lbs. (953 g.). Because of the shape of the frame, this Rack is very strong.

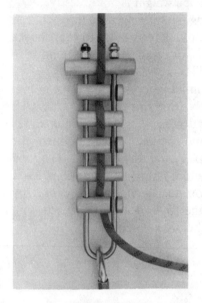 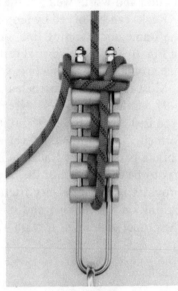

Fig. 6-98   **Fig. 6-99**

The RSI Rack features a fixed top bar, and fixed #3 and #5 bars. These bars will slide on the frame, but will not open. This design, along with the straight cut slots in bars #2, #4, and #6, makes for increased safety when rigging. Rigging is done by simply opening the hinged bars, one at a time, and inserting a loop of rope around them. Fig. 6-98 shows the Rack properly rigged for a single line rappel.

Friction in this Rack is varied by pulling down on the braking end of the rope, as in most other Racks and descenders, or by pushing the brake-bars upward on the frame. The number of bars in contact with the rope can be changed while on rappel, but the effort required to make the change can vary from clumsy to impossible, depending upon the weight of the rope.

Fig. 6-99 shows one of several methods for locking a rappel when using a RSI Rack. The extended length of the top bar makes this an easy procedure, and even more secure variations are possible. The extra metal in the top bar also enables it to dissipate heat quite well.

Replacing the bars on a RSI Rack is easy. Simply remove the cap and lock nuts from the frame, remove the worn bars, and slip on new ones. Be sure the bars are replaced in proper order, with the slots facing the correct direction. Replace the nuts properly – without overtightening them. Before every use, check by hand to make sure the nuts are secure. This advice applies to any Rack which uses lock nuts to retain the brake-bars.

Because of the amount of metal in most Racks, they dissipate heat very well, which makes them excellent for long rappels. In order to increase dissipation efficiency even more, Racks can be equipped with hollow aluminum, steel, or stainless brake-bars (Fig. 6-100). Steel and stainless bars last longer than aluminum ones, and don't discolor the rope; they also dissipate heat better than most solid aluminum bars. However, they don't provide as much friction. When rappelling on steel bars, you may have to add more bars in order to obtain sufficient friction. If you are accustomed to rappelling on aluminum bars – be careful!

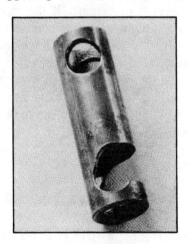

Fig. 6-100

Brake-bars – particularly hollow ones – should be replaced before worn excessively. The sharp edges on a worn-thru hollow bar can cut a rope. A very worn solid bar is prone to breaking. When replacing bars, be sure they are put on in the right order – and with the slots facing in opposite directions. Replace the nut without overtightening it. Never rappel on a Rack without first checking the condition of this nut.

When doing rappels with a Rack, its sometimes handy to regulate speed and obtain extra friction by wrapping the rope around your leg, Tarzan style, as shown in Fig. 4-5. This can save some wear on the hands on a long rappel. Just be careful!

When rappelling on a Rack, be extra careful if you have long hair or a long beard. The great length of these descenders makes them apt to grab such things. Always carry prusik slings for doing self-rescue, and a knife, in case something must be cut. The length of a Rack also complicates crawling over edges, so be careful about getting hung.

Although bottom belays are advisable in most rappel situations, be aware that they may be useless – or may cause an accident – during a Rack rappel. If a Rack is rigged with little friction to begin with, and the rappeller loses control and cannot squeeze the brake-bars together, pulling on the braking end of the rope may have little effect – particularly if the rappeller is several hundred feet above ground. Rope stretch, combined with the inherent spreading of the bars on the Rack frame, may make braking impossible. One rappeller was killed because of this effect – even though the belayer applied his total weight to the rope! As stated earlier, bottom belays can also cause accidents if the rappeller is pulling upward on the braking line and the belayer is pulling down.

RAPPELEVATOR

The Rappelevator (Fig. 6-101) is a very old and unique descender. It is formed from a bar of steel, and has an angled slot in one side. The slot permits a rope to be inserted into the descender. The small hole near the slot is for attachment of a harness carabiner.

The Rappelevator is rigged and used in a manner similar to a Carabiner Wrap. Rigged correctly, the rope should not slip out of the slot. Caution must be observed, however.

Fig. 6-101

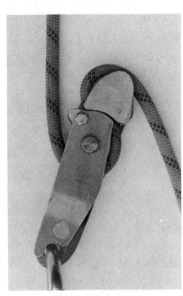

Fig. 6-102

DRESSLER

The Dressler Descender, by Brevete, is shown in Fig. 6-102. It is made of aluminum, and features an integral "dead-man" lever. The rope is inserted into this descender by removing the harness carabiner from its hole, and rotating the "dead-man" lever counterclockwise. Once the rope is in place, the lever is returned to the position shown in the photo, and the carabiner inserted into its hole.

The Dressler, like the Petzl STOP, has a spool attached to the "dead-man" lever. In use, rope friction around this spool tends to rotate the lever clockwise, squeezing the rope between a tooth on the spool and the upper rope guide. By squeezing on the lever with the balance hand, this descender can be used like any other, with braking accomplished by pulling on the braking end of the rope. During an emergency, if the balance hand lets go of the lever, the rappel will be instantly halted.

Because the rope merely lays over the top rope guide, great care must be used to keep it in this position. If it should fall to the side, braking action could be altered, or the device could jam. In general, this descender should be used in the same manner as a STOP, but with much greater caution.

ESCAPELINE

The Escapeline, by Mar-Mex International, is a unique device which has features of a Pressure Plate and a conventional descender. It is made in two sizes, as shown in Fig. 6-103. The small unit weighs 6 oz. (165 g.) and can be used with ropes ranging in size from 3/8 inch (9.5 mm) to 7/16 inch (11 mm). The large unit weighs 13 oz. (376 g.) and can be used with ropes from 7/16 inch to 1/2 inch (13 mm).

Escapelines are made of an anodized 6061-T6 aluminum extrusion, which has properties similar to aluminum plate. The two halves, seen in Fig. 6-104, are connected by a hinge joint at the rear. The knob is secured onto a threaded stainless stud, which is held in place by a stainless pin. Knobs are available in aluminum, or in Bakelite plastic with a threaded brass insert. This insert is wide enough so that it cannot pull thru the slot in the rear plate. The large Bakelite knob is easier to grip than the smaller aluminum one, particularly when wearing gloves.

Escapelines are supplied with a double-braided polyester rope, which has a loop spliced into one end. They are available in a kit which contains a seat harness, carabiner, and rope bag.

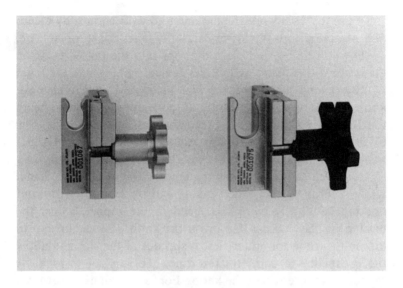

Fig. 6-103

Fig. 6-104

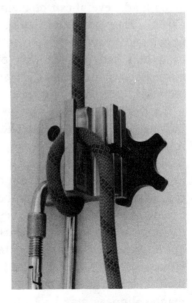

Fig. 6-105

To use the Escapeline, first attach it to your harness by means of a carabiner thru the lower carabiner hole. Next, open it up, as shown in 6-104, and place the rope into the slotted track on the inside of the plates. Close the device, making sure the knob is fully behind the lip on the rear plate, and then tighten the knob as hard as possible. Bring the rope upward, from beneath the Escapeline, and place it into the notch on the side (Fig. 6-105). Make sure the rope is securely in this notch. Do not rappel with the rope resting only in the track.

To rappel, carefully loosen the knob with your balance hand, while securely holding the braking end of the rope with your braking hand. To vary the speed of the rappel, adjust the tension on the braking line or on the knob. Be careful not to loosen the knob too much, lest it slip out of the slot and allow the Escapeline to come off the rope. To stop, or to lock the rappel, simply retighten the knob. For safety, an extra turn can be taken over the notch's horn when locking off.

If one Escapeline cannot provide sufficient friction to lower a given load, several can be connected in series, as shown in Fig. 6-106. They are coupled together by carabiners thru both the upper and lower carabiner holes. When used in this manner, each unit can provide part of the friction, or they can be alternately tightened and loosened on the rope. This latter method allows some to cool while others are bearing the load.

Besides being used as rope brakes, Escapelines can be used for locking off ropes in rescue and raising systems. Tension on a rope can be released by simply turning the knob, instead of having to first unweight the rope, as with most locking devices and ascenders. However, caution must be observed when using the descender in this manner, because it may not lock tightly on certain rope diameters and types.

Escapelines operate most satisfactorily with the rope provided, but under some circumstances they can be used with other types. However, they must never be used with ropes larger or smaller than specified, or an accident could result. With certain ropes, mantel slippage may be a problem. It may also be difficult to get the plates closed when using very stiff ropes.

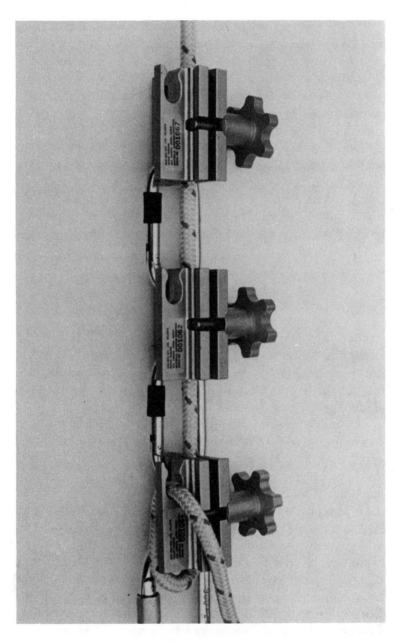

Fig. 6-106

Because of their weight, Escapelines can dissipate heat fairly well. However, they have been known to get very hot when used on certain static ropes, so should never be used for fast rappelling – except in an emergency.

As stated earlier, this descender must never be simply clamped onto the rope like a Pressure Plate. Although the rope track keeps the rope fairly round and does not crush it, as occurs in certain Pressure Plates, variations in rope thickness and surface condition make such rappelling very dangerous. Hitting a slick section on a rope, or one of lower diameter, could cause loss of control. The clamping pressure may also cause melting of the rope's surface. Always rappel with the braking end of the rope properly placed in the side notch.

TRITON

The Triton (Fig. 6-107) is made by Forrest Mountaineering, and is a combination rappel, belay, and climbing device. When not being used for rappelling or belaying, it can function as a nut, similar to a #8 Forrest Titon. Tritons are made of forged aluminum and are anodized. They weigh 4 oz. (111 g.).

For single-line rappelling or belaying, the Triton is rigged as shown in Fig. 6-108. A loop of rope is pushed thru one of the upper slots, and then a carabiner is clipped thru this loop and the large eye in the stem. In operation, the device functions much like a Sticht Plate, and braking is achieved by pulling on the braking end of the rope. Fig. 6-109 shows the Triton rigged for a double-line rappel.

The eye of the Triton is large enough to accommodate two carabiners, and its prime function is to keep the device from sliding down the rope during belaying. It also keeps the carabiner from jamming against the slotted plate, which helps the Triton work more smoothly than most Sticht Plates. Because of the amount of metal in Tritons, and the amount of surface area, they dissipate heat better than most small belay plates.

Fig. 6-107

Fig. 6-108   **Fig. 6-109**

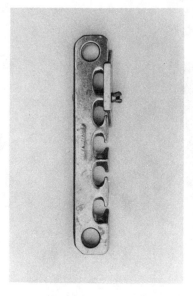 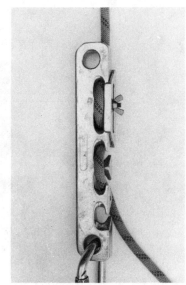

Fig. 6-110   **Fig. 6-111**

WHALE'S TAIL

The Whale's Tail (Fig. 6-110) is one of the older descender designs. It is machined or cast from aluminum, and is used for single-line rappelling. As can be seen in Fig. 6-111, a Whale's Tail is rigged by zig-zagging the rope thru the triangular studs on the side and then securing it in place with the lock plate. Never rappel without this lock plate firmly in place, and don't rappel on oversized ropes which might deform the plate. During the rappel, make sure the wing nut does not loosen.

Some early model Whale's Tails were designed without lock plates. The rope was held in place only by the ears on the studs. These models should be used with extreme caution. With any type of Whale's Tail, friction can be varied during the rappel by taking the braking end of the rope around, or removing it from contact with, one or more of the lower studs.

MILLER DESCENT DEVICE

The Miller descender, shown in Fig. 6-112, is manufactured by the Miller Equipment Company. It weighs 1.75 lbs. (794 g.), and comes with a ½ in. (13mm) braided polyester rope, which has end loops formed by the Thimble/Swage Sleeve method. Attached to the rope is a heavy tubular protector.

Similar to the Sky Genie, the Miller descender consists of a cast aluminum spindle (Fig. 6-113) and a heavy tubular cover made of anodized sheet aluminum (Fig. 6-114). Unlike the Sky Genie, this descender has spiraled rope guides on the spindle. These guides provide extra amounts of friction and improve heat dissipation, but their presence renders the device capable of single-line rappels only.

To remove the cover from the descender, the thumb nut is loosened, and a spring plunger near the bottom is depressed. The cover can now be slipped downward and off the spindle. The rope is then placed in the upper rope guide and wound around the spindle an appropriate number of turns. Fig. 6-115 shows the descender rigged for maximum friction. Instructions for rigging are printed on the cover. Once the rope is in place, the cover is slipped on, the thumb nut is tightened, and a harness carabiner is clipped into the lower eye (Fig. 6-116).

Once on rappel, the rigging of this descender cannot be changed. The most practical way to increase friction, when needed, is to wrap the braking end of the rope around the horn projecting from the lower end of the descender (Fig. 6-117). Locking the rappel can be achieved by wrapping the rope around the horn several times, or securing it with a half hitch.

Like the Sky Genie, the Miller descender cannot be rigged without first removing the harness carabiner – and it must never be used without the cover properly in place and tightly secured. When rigging, be careful not to drop the cover – and make certain that the rope enters and exits thru the proper guides – not thru the slot in the side of the cover. This descender should be used only with the rope provided, and the rope should be replaced immediately after hard usage.

Fig. 6-112

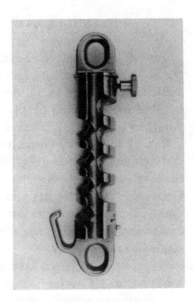

Fig. 6-113

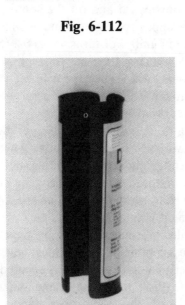

Fig. 6-114

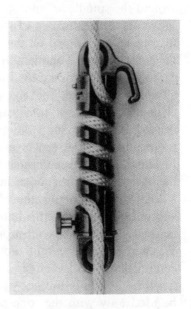

Fig. 6-115

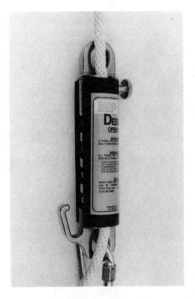

Fig. 6-116   Fig. 6-117

HALL SIDEWINDER

Made by Lirakis, the Sidewinder is formed of stainless rod and then welded. It is available in two sizes (Fig. 6-118), with the small unit weighing 4.2 oz. (119 g.) and the large weighing 1 lb. (450 g.). It is rigged, as shown in Fig. 6-119, by simply pushing the rope thru the middle and looping it over the horns.

The large Sidewinder can be used with ropes of up to ⅝ in. (16mm) diameter, and the small one with ropes up to ½ inch. The small unit usually provides more friction than the large one, because of the sharper bends in the rope, but extra friction can be obtained in either by "double wrapping" (see Chapter 13, "Extra Friction"). Unlike the MSR Longhorn, extra friction cannot be obtained by taking an extra wrap around the horns, because this normally causes the descender to lock. When actual locking is desired, several wraps should be used (Fig. 6-120).

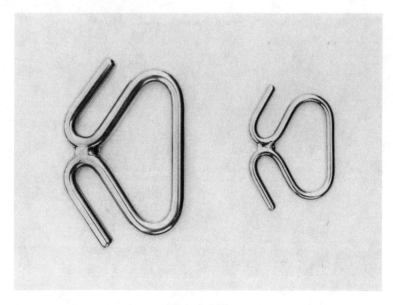

Fig. 6-118

Fig. 6-119

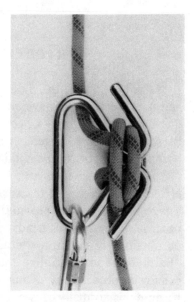

Fig. 6-120

LEWIS AUTOLOCK

Made by Lewis, this aluminum descender operates differently than most autostop devices. As can be seen in Fig. 6-121, the descender has a body which does not rotate open. To insert the rope, the "dead-man" lever is squeezed, the rope is pushed thru the body, and then looped over the spool on the lever. Fig. 6-122 shows the descender rigged for double-line rappelling.

 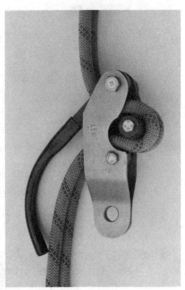

Fig. 6-121   **Fig. 6-122**

Automatic braking in this descender occurs because the rope is pinched between the lever and body spools. When the "dead-man" lever is squeezed, more clearance is created, and the rope can slide. Although it is possible to descend by regulating tension on the lever, it is best to keep the braking hand on the braking end of the rope, and descend in a normal fashion. Sudden braking with the lever should be reserved for emergency stops. When rappelling with this descender, be careful not to hit against objects which might pull the rope off one of the lever spools.

Fig. 6-123   **Fig. 6-124**

COE

The Coe descender, by Penticton Engineering, is shown in Fig. 6-123. It resembles a Brake-bar and Carabiner rig, but in fact the "biners" are solid, with no gate, and the brake-bars are permanently attached. The descender is made in two models, one with a short brake-bar, and one with a long brake-bar. The purpose of the long bar is for locking the rappel (Fig. 6-124).

A single Coe descender can be used for rappelling, but for adequate friction, it is best to couple two or more units together, using carabiners or lock links. This descender is rigged and operated like a standard Brake-bar/Carabiner descender. The bars are flipped open, and the rope is inserted behind them. Safety is high because of the straight slots in the brake-bars, and because there are no carabiner gates to break away during heavy loading.

AMERICAN

The American Descender (Fig. 6-125), by American Rescue Systems, is designed for lowering heavy loads on ropes from 7/16 in. (11mm) to 5/8 in. (16mm) in diameter. It is machined from aluminum and held together with stainless bolts and fittings. It weighs 3.24 lbs. (1.47 kg.). It is also designed to be used as a high strength pulley system.

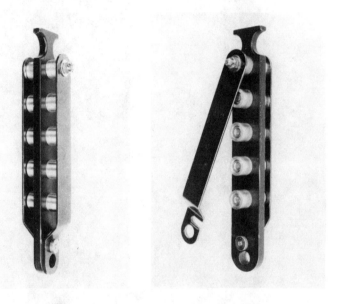

Fig. 6-125   Fig. 6-126

As can be seen in Fig. 6-126, the side plates swing open to expose the spools. Single-line rigging is accomplished by placing a rope against the upper spool and then zig-zagging it downward thru the lower ones (Fig. 6-127), in a fashion similar to rigging a Rack. Once rigged, the side plate is closed and secured with the harness carabiner. This descender cannot be opened with the carabiner in place. Double-line rappelling is accomplished by opening both side plates and rigging both sides of the descender, as shown in Fig. 6-128.

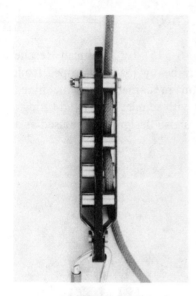

Fig. 6-127

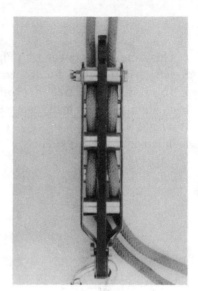

Fig. 6-128

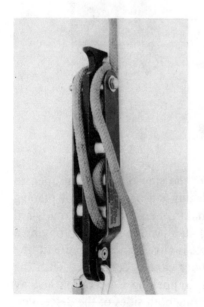

Fig. 6-129

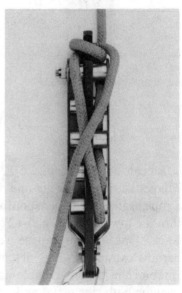

Fig. 6-130

The amount of friction available in this descender is determined primarily by the number of spools in contact with the rope. Adequate amounts of friction can usually be obtained when using large rescue ropes, particularly when rappelling double-line. Once rigged and on rappel, friction cannot be decreased; but it can be increased, on single-line rappels, by looping the braking end of the rope around the horn at the top of the descender (Fig. 6-129). Locking the rappel is accomplished by wrapping the rope around the horn several times (Fig. 6-130).

The two upper spools on the American descender are not rigidly bolted to the frame. They are free to rotate, so that the descender can be used as a pulley. This can be of great value in rescue operations such as Tyrolean traverses. The lower spools are held in place by strong stainless bolts, and they should be checked often to insure tightness. For the descender to operate properly and safely, these spools must never rotate.

MUNTER

Manufactured by Eiselin-Sport, this aluminum descender (Fig. 6-131) is designed to be used in several ways. Fig. 6-132 shows it rigged like a Figure-8 descender, with the harness carabiner attached thru the large hole at the bottom. The ears projecting from upper ring help prevent the rope from slipping upward and cinching, as occurs with normal Figure-8's. Fig. 6-133 shows a similar rigging which provides lower amounts of friction. This mode of rigging totally prevents the rope from slipping upward and cinching. For double-line rappelling, it can be rigged in a manner similar to that shown in Fig. 6-52.

When it is important that the descender not cause twisting and kinking in the rope, such as when rappelling on laid rope, the Munter can be inverted and rigged as shown in Fig. 6-134. Here, a carabiner is used as a brake-bar, and the harness carabiner is attached to the large eye. Using two carabiners in parallel will increase the friction available.

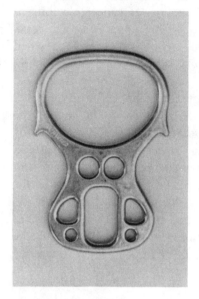

Fig. 6-131

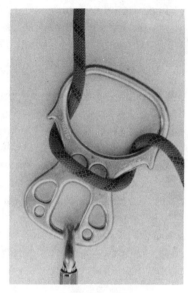

Fig. 6-132

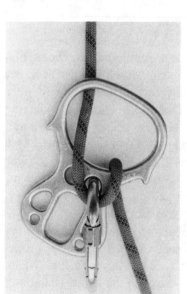

Fig. 6-133

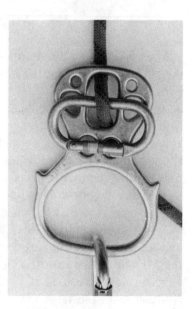

Fig. 6-134

Thank you for your purchase!

This is a quality book by SEARCH. Great amounts of time and money have been spent to make it physically durable and technically correct. In the event that significant changes or corrections become necessary, we would like to have your name in our computer so that we can make you aware of these changes. Please fill out this card and return it to us. The blank lines near the bottom may be used to list comments about this book. This will help us improve future editions. Thank you.

_____/_____
Name   Organization

_____
Mailing Address

_____
Where was this book purchased?

_____

_____

_____

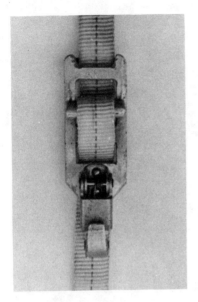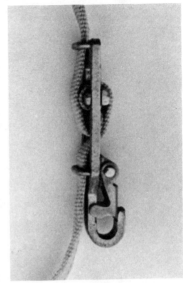

Fig. 6-135

PERSONNEL LOWERING DEVICE

This military descender is made of steel, and is designed for rappelling on nylon webbing. Fig. 6-135 shows the front and side views of the descender, properly rigged. Braking is effected by pulling on the braking end of the webbing, as with most descenders. In normal use, the webbing is faked into a flat pouch, and the descender rigged and ready for emergency use.

LATOK TUBER

This unique descender, by Lowe Alpine Systems, was designed as a belay device, but can also be used for rappelling. It is machined from aluminum and has an attached bail, for securing

Fig. 6-136

Fig. 6-137

it to a carabiner; in the same way that a lanyard is used to hold a Sticht Plate near the carabiner. The Tuber (Fig. 6-136) weighs 2 oz. (57 g.), and has fins for dissipating heat.

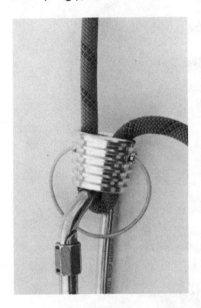

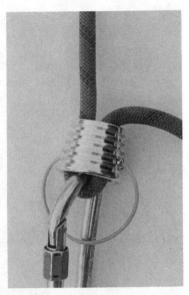

Fig. 6-138   Fig. 6-139

The Tuber is rigged much like a Sticht Plate. One or two ropes are pushed thru the device, and then clipped into the carabiner. For maximum friction, the small end of the Tuber is placed against the carabiner (Fig. 6-138). This mode is most often used with small diameter or very flexible ropes. When less friction is needed, such as when using large or stiff ropes, the big end can be placed against the carabiner (Fig. 6-139). Fig. 6-137 shows a Tuber being used for double-line rappelling.

Tubers are designed for use with ropes from 3/8 in. (9mm) to 7/16 in. (11mm) in diameter. However, present models may not work well with old and dirty 11mm ropes, because of the increased rope diameter. In use, Tubers generally work very smoothly and provide great amounts of friction. Their fins help to dissipate heat, but they can get very hot on long, fast rappels.

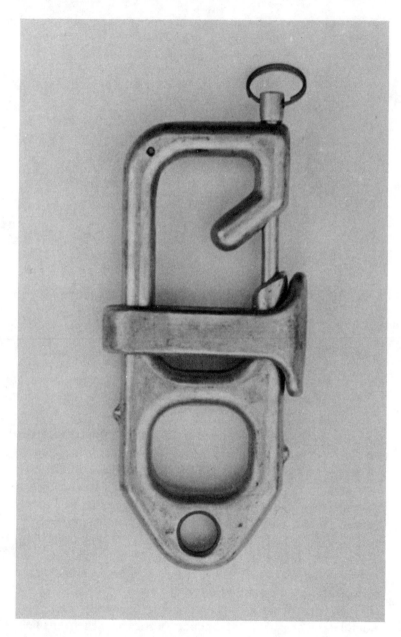

Fig. 6-140

HOBBS HOOK

Made by Bry-Dan Corp., this aluminum descender can be rigged in many ways. Fig. 6-140 shows the Hobbs Hook with its movable brake-bar in place. The small hole, at the bottom of the descender, is for attaching a harness carabiner.

Fig. 6-141 shows the Hobbs rigged in a fashion similar to a Carabiner Wrap. The gate, in the right side of the descender, is slipped open by pulling upward on the ring at the top. It is spring loaded so that it closes when the ring is released. The amount of friction available in this mode is determined mainly by the number of times the rope is wrapped around the frame.

Fig. 6-142 shows the Hobbs rigged like a Brake-bar/Carabiner descender. The brake-bar slides entirely over the frame, and cannot come off once the device is rigged for rappelling. Because the metal in the brake-bar is fairly thin, it will wear thru quickly, particularly when rappelling on dirty ropes. It should be replaced at the earliest sign of wear. Rigging the descender as shown in Fig. 6-142 causes only minimal twisting of the rope below the rappeller; whereas the rigging shown in Fig. 6-141 causes the same twisting and kinking problems that are experienced with Brake-bar and Carabiner descenders.

If more friction is required than is available in Fig. 6-142, the Hobbs Hook can be rigged as shown in Fig. 6-143. Here, a carabiner is used as a brake-bar, and essentially doubles the amount of friction available. The rope must only run across the backside of this carabiner – it must never run across the gate.

A unique method of rigging is shown in Fig. 6-144. The rope is pushed thru the lower eye and looped over the gate horn. This method helps eliminate kinking in the rope. If more friction is needed, the rope can also be wrapped around the frame several times, similar to the rigging shown in Fig. 6-141.

The Hobbs Hook can also be rigged like a Figure-8 by simply inserting a loop of rope thru the lower eye and clipping it thru the harness carabiner. However, there is seldom any practical reason for doing so. Although only shown rigged single-line, the same techniques can be used for double-line rappelling.

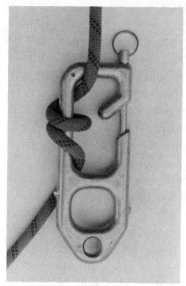

Fig. 6-141

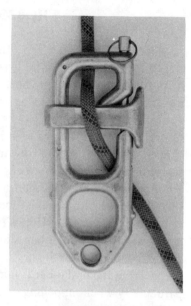

Fig. 6-142

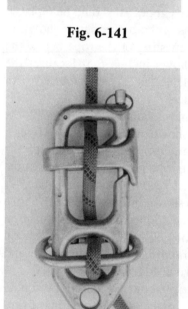

Fig. 6-143

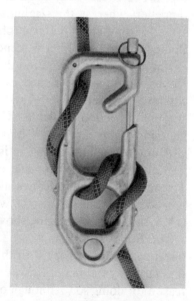

Fig. 6-144

FAMAU

The Famau (Fig. 6-145) is made of aluminum, and is rigged like a Figure-8 (Fig. 6-146). Rappels can be locked by taking the braking end of the rope down around one of the lower ears and up and over the horn projecting from the center eye. The carabiner hole in the Famau is small, and will not accommodate most of the large modern carabiners.

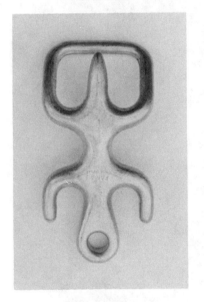 

Fig. 6-145   **Fig. 6-146**

BATBRAKE

The Batbrake (Fig. 6-147) is a belay device manufactured by Edelrid, which can also be used for rappelling. It is made of aluminum and weighs 1.9 oz. (54g.). Fig. 6-148 shows it rigged similar to a Figure-8 descender. This provides moderate amounts of friction, but normally not enough for safe rappelling.

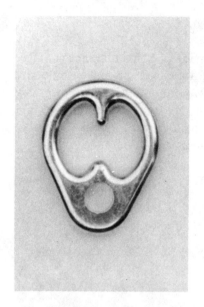

Fig. 6-147

Fig. 6-148

Fig. 6-149

Fig. 6-150

Fig. 6-149 shows how more friction can be obtained by taking the rope thru the attached carabiner. For belaying, a large carabiner is used, so that the Batbrake can rotate back thru the carabiner when it is necessary to change the direction of pull on the rope. Used in this manner, it functions like a Munter Hitch. Fig. 6-150 shows how the device may be used for double-line rappelling or belaying. Because of weight – and available friction – Batbrakes should be used with caution.

PRO-PAK

The Pro-Pak descender, shown in Fig. 6-151, is made by Rappel Rescue Systems, Inc. It is forged of 6061-T6 aluminum and anodized, and weighs 9 oz. (260 g.). It is designed for use with Kevlar webbing, and comes in a kit consisting of a harness, carabiners, and up to 1,000 ft. of webbing in a belt pouch.

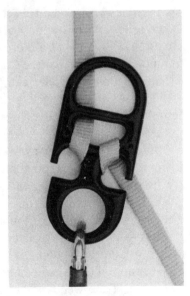

Fig. 6-151   **Fig. 6-152**

The Pro-Pak descender is designed for use with 5/8 inch tubular Kevlar webbing, which tests at over 9,000 lbs. (4,082 kg.). This webbing is pure Kevlar, and not covered by a mantel, so that it is easy to inspect for abrasion and wear. It is provided with a sewn attachment loop at the end, so that strength is not sacrificed by using a knot when anchoring. It comes equipped with a special edge protector which is made of heavy nylon webbing and stainless steel strips. The protector is sandwiched around the webbing, but is free to slide, so that it can be placed over sharp edges to protect the webbing from cuts.

Kevlar has certain advantages over nylon in certain rappel applications. For all practical purposes it does not stretch, which improves safety when going over edges (see Chapter 11). However, this lack of stretch means that the energy generated by fast rappel stops is transmitted to the anchor and to the rappeller's body. When rappelling on Kevlar, it is important to rappel smoothly, and to use strong anchors.

Kevlar does not cut as easily as nylon, and it withstands much higher temperatures. Over hot edges, or in direct flame, Kevlar has been shown to retain its strength several times longer than nylon rope. This could prove crucial in certain fire rescue situations. Kevlar is also more resistant to chemicals than other common rope materials (see Chapter 3).

The Pro-Pak descender can be rigged in several ways. Fig. 6-152 shows the appropriate rigging when minimum friction is desired. The method shown in Fig. 6-153 can be used when greater amounts of friction are needed. By pushing a loop of webbing thru the upper eye (above the cross bar), bringing it down and over the carabiner attachment eye, pushing it thru the eye below the cross bar, and once again looping it over the carabiner eye, maximum friction is available (Fig. 6-154). This latter rigging is similar to "double wrapping" a Figure-8.

The Pro-Pak descender is controlled like a Figure-8, but it is shaped so that the webbing is not bent sharply. This helps limit self-abrasion of the Kevlar fibers, and increases the strength and useful life of the webbing. In critical situations, however, this webbing is designed to be used one time only.

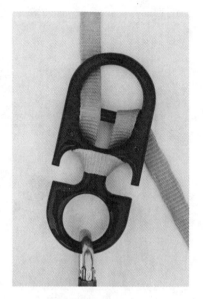 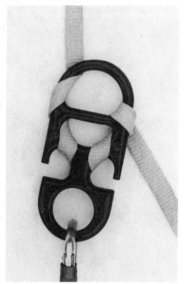

Fig. 6-153   Fig. 6-154

FIRE CROSS

The Fire Cross, by Colorado Mountain Industries, is a versatile belay/rappel plate which is designed for use with 7/16 to 1/2 inch rescue rope. Shown in Fig. 6-155, it is forged of aluminum, anodized, and weighs 4 oz. (116 g.).

As can be seen in Fig. 6-156, the Fire Cross is rigged like a Sticht Plate. When extra friction is needed, the braking line can be pulled into one of the slots in the plate (Fig. 6-157). If slightly less friction is needed, the rope can be taken over the plate and into the slot on the opposite side.

When great amounts of friction are needed, the rope can be taken thru several slots, as shown in Fig. 6-158. This method can also be used to lock the rappel. For more secure locking, a half-twist loop can be formed in the braking line, and then looped into two of the slots in the plate, creating a secure hitch.

Fig. 6-155

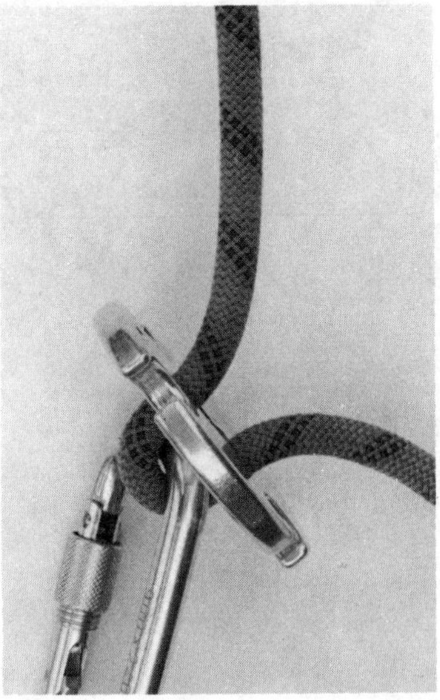

Fig. 6-156

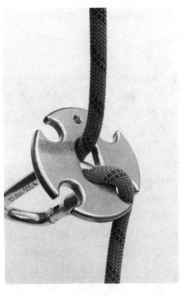

Fig. 6-157

Fig. 6-158

Fig. 6-159 shows this hitch in place. This method of locking is easily accomplished with one hand. To unlock, simply push the braking line back thru the hitch, in order to loosen it, and then remove the line from around the "ear" on the plate.

Fig. 6-159

ET CETERA

CARABINERS

Also called Karabiners or Snap Links, these devices are made of aluminum or steel and have a spring-loaded gate which allows them to be clipped onto ropes, descenders, and etc. The pin in the upper end of the gate slips into a notch when the gate is closed, enabling both sides of the carabiner to support weight (Fig. 7-1).

Fig. 7-1

Some carabiners have a threaded sleeve on the gate which can be screwed upward to prevent the gate from opening accidentally. Others have gates which lock automatically upon closure. These Locking Carabiners (Fig. 7-2), are used whenever the ultimate in security is required. Two regular carabiners can be used in place of a Locking by orienting them so their gates are on opposite sides or facing in opposite directions (Fig. 7-3).

Carabiners come in a wide variety of shapes and sizes, but the most common are D's (Fig. 7-4) and ovals (Fig. 7-5). There is a variation of the carabiner called a Snaphook (Fig. 7-6). It is sometimes used at the end of a rope to provide quick hookups.

The gate side is normally the weakest side of a carabiner. D's, because of their shape, place only a small part of the applied load on their gates. This allows them to support almost twice as much weight as ovals. Because of this superior strength, D's should always be used, unless the oval shape is required in order to make a rigging operation work more smoothly.

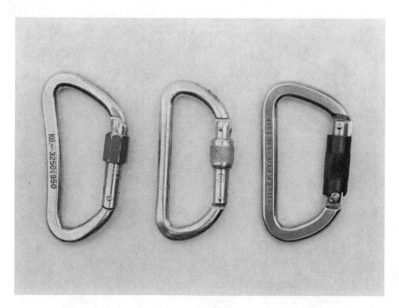

Fig. 7-2

Fig. 7-3

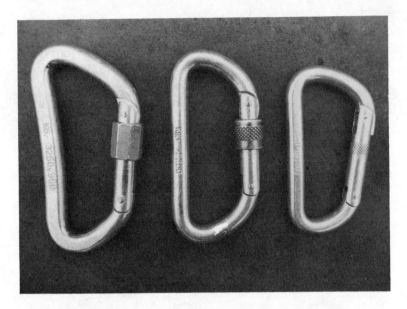

Fig. 7-4

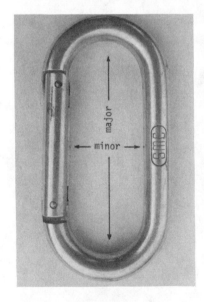

Fig. 7-5

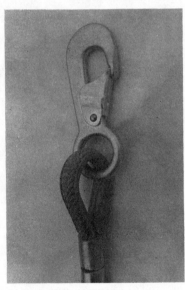

Fig. 7-6

Older types of carabiners were made of mild steel, and will seldom support over 3,000 lbs. (1,361 kg.). Although countless rappels have been done on these type carabiners, they should not be used; they have too little margin of safety. Modern aluminum carabiners are much stronger, with strengths ranging from 4,560 lbs. (2,068 kg.) to over 7,000 lbs. (3,175 kg.).

A quality U.I.A.A. approved aluminum carabiner will typically hold over 7,000 lbs. along its major axis (with the gate closed), and over 2,600 lbs. (1,179 kg.) with the gate open. It can also support a load of 2,094 pounds (950 kg.) across the minor axis. Aluminum harness carabiners should preferably be U.I.A.A. approved types (the U.I.A.A. symbol ⓤ will be stamped on the side), or other high quality type that will hold at least 6,000 lbs. (2,722 kg.). All harness carabiners should be of the locking type.

When extremely high strength is needed, the newer type steel and stainless carabiners should be used. They come in all shapes and styles, and some will support up to 16,000 lbs. (7,258 kg.). These steel varieties also come in large models, for hooking over litter frames and ladder rungs. Some are big enough to put your fist thru. When weight is not of paramount importance, these steel varieties are the carabiners of choice.

As shown in Chapter 6, the end tabs on carabiner gates cannot hold much weight. If side pressure is applied to a gate, the tabs will typically snap at forces of only 500 to 2,500 lbs. (227 to 1,134 kg.). Carabiner gates must never be heavily stressed!

Another device, which is often used in place of a carabiner, is shown in Fig. 7-7. It's called a Lock Link or Maillon, and the threaded nut serves both as gate and lock. This type "gate" must be screwed open and closed, and if grit gets into the threads, great difficulties can result. They are available in steel or aluminum and are as strong or stronger than most carabiners.

Carabiners do age and do get weak from repeated stress. When a carabiner shows much wear, it may be time to retire it. This applies especially to harness carabiners. Never hang your life on anything which is badly worn or damaged – your life is worth more than that. Also, in order to prevent carabiner gates and locks from jamming with dirt and grit – never oil them.

Fig. 7-7

WEBBING

Webbing, also called Tape or Flat Rope, comes in solid and tubular forms (Fig. 7-8), and is used for harnesses, slings, rope protectors, and etc. Solid webbing is like seat belt webbing, whereas tubular is woven into a hollow tube. In an emergency webbing can be used as a rappel rope (with some descenders).

Fig. 7-8

Webbing used by rappellers should be made of either nylon or polyester (Dacron). Nylon is usually best, but if long exposure to sun is probable, polyester is the way to go. Although webbing is strong (1 × 1/8 in. tubular nylon will hold 4,000 lbs. (1,814 kg.), it cannot absorb shock like a rope. One-inch webbing has been known to break from the force of a man falling only four feet. If you use webbing in your anchor system – rappel gently. Also, remember that webbing loses its strength much faster than rope. It wears especially fast along the edges. Only use webbing which is strong, new, and in good condition!

When tying pieces of webbing together, bear in mind that many knots will not hold well. The safest knots for this purpose are Water knots and Double Fisherman's knots. Generally, the best way to secure webbing is by sewing. See Chapter 5 for more info.

ROPE PROTECTOR

When ever a rope runs across a rough or sharp surface, a protector should be used to prevent rope abrasion and damage. Fig. 7-9 shows a protector doing its job on a sharp edge. The best protectors are made from 2 to 3 foot lengths of flexible tubular material, like garden hose or tubular nylon webbing, and should be of sufficient diameter so that the rope can be easily pushed thru.   Garden hose works fine, particularly if it is fabric reinforced. Its disadvantages are bulk and a tendency to slide down the rope if not secured in place. One-inch tubular nylon webbing makes the most practical protector. Its very flexible and does not readily slide down the rope. Since it does not have tension on it during use, it can't be cut easily by a sharp edge. Don't melt the ends of the webbing – leave them frayed. Melted ends crack and the sharp edges will pull fibers out of the rope.

Never use tubing which is split down one side. This type protector is easy to put on the rope, but the rope will roll out of it if there is any side movement. Most places where split-side protectors can be safely used don't require protectors at all!

Fig. 7-9

Besides protecting rope from cuts, protectors actually increase rope strength by increasing the radius of the edge (see Chapter 3). The proper use of protectors is described in Chapter 11.

The main disadvantage to protectors becomes obvious when you try to go back up your rope. Tubular nylon protectors can be climbed with ascending knots and some mechanical ascenders. Protectors made of hose cannot be climbed, but must be crossed. In either case, so long as the length of the protector is short, you can ascend the rope with the proper equipment. Chapter 12 has specific information on these ascending methods.

GLOVES

Never do rappels without a good pair of gloves. Doing so could cost you your life! On a long, fast rappel, even heavy leather gloves can be burned thru when trying to stop quickly. Without gloves, the rope would be rubbing against bone!

Leather and Kevlar are the best materials for gloves, although cotton or wool will do in a pinch. Never use gloves of normal plastics or synthetics; rope friction can melt them in seconds.

DESCENDING RINGS

Also called Abseil Rings, these little devices expedite rope retrieval (see Chapter 13, "Rope Retrieval"), and can be used to build special gear. They are made of steel or aluminum and are usually 1.5 to 2.5 inches in diameter. The most popular type is made of aluminum sheet which has been rolled into a hollow tube (Fig. 7-10). This type has no weld joint, is lightweight, yet will hold several thousand pounds.

Fig. 7-10

Descending rings are very versatile and quite inexpensive. They keep slings from being cut, and make pulling a rope down much easier, particularly if its wet. Putting two rings side-by-side makes retrieval even easier, and doubles the margin of safety.

KNIFE

A knife can't be used for rappelling, but it is a useful item to carry. It can be used to cut jammed lines or fend off wild beasts. It should be a fixed blade type (Fig. 7-11), or something which can be opened with one hand. In most cases scissors can be used

Fig. 7-11

instead, and could prove much safer. The main reason for a cutting implement is in case hair or clothing gets caught in a descender. It should be used only as a last resort, after safer methods of unjamming have proved futile (see Chapter 13, "Getting Unjammed"). Great care should be taken during the cutting process. As one girl was cutting her hair, she severed her rope!

FIFFI HOOK

This device (Fig. 7-12) is made of heavy steel, and can be used at the end of a rope to expedite retrieval. This is described in Chapter 13 under "Rope Retrieval." Because Fiffis have been known to break, use only the strongest ones for rappel hookups. If there is any doubt about strength, tie two or more together so that they function as one – or better yet, use another method of retrieval.

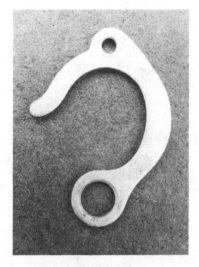

Fig. 7-12   Fig. 7-13

ETRIER

Pronounced "a'-tree-a," this device is a short rope ladder used in technical climbing. Some etrier have three or four little aluminum rungs, but most are simply made of webbing or cord. Etrier are very useful in rappelling when one has to go off an overhang. A lightweight etrier made of thin cord (Fig. 7-13) will fit in a pants pocket, and could be a lifesaver.

HELMET

While rappelling you can hit your head on things, or things can hit you on the head. Once you've had a boulder from nowhere whiz past your head, you will be forever convinced of the value of a helmet. Helmets are also good for digging foxholes, carrying rocks, and wearing to formal social events.

The best helmets are molded from ballistic nylon (Kevlar), or polycarbonate resin (Lexan). Fiberglass and metal helmets (tho not as strong) are also usable. The cushioning – the part of the helmet which absorbs energy – should meet all the latest specifications for energy absorption. The suspension and chin strap should also be of an approved sort.

For special missions and for training purposes, a full face shield should be attached to the helmet (Fig. 7-14). Bubble-type face shields are great for keeping limbs out of your eyes on night rappels. During training sessions, if limbs aren't a problem, a football helmet with face guard could be used.

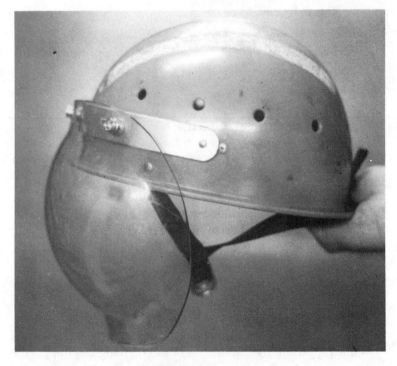

Fig. 7-14

KNOTS

A rappeller's life depends on many factors, not the least of which is how well he ties knots. The knots shown in this chapter are those most commonly used in rappelling. You must learn them so well that you can tie them in the dark while in a cold shower! Some other knots are shown in other chapters. They are for special purposes and you should learn them too. First, however, learn the ones in this chapter – you will be using them constantly. When you are studying, remember that a knot tied wrong could result in death.

Because great strain is placed on rope fibers by the sharp bends in a knot, ropes usually break at or near knots (unless they get cut). Different knots put different amounts of stress on different types of rope. In other words, a knot which may be strong in laid rope may not be so strong when used in a kernmantel rope or webbing. Also, certain knots do not stay tied well in certain types of rope and webbing. The knots shown here are a compromise of strength, simplicity, and ease of untying. All can be used with rope, but only the Water knot and the Double Fisherman's should be used to join two pieces of webbing.

After mastering these basic knots, you would do well to study books on knots for mountaineering. They contain data on knot strength for various type rope, and show a much wider variety of knots than shown here.

OVERHAND KNOT
(Fig. 8-1)

Fig. 8-1

FIGURE-8 LOOP
(Fig. 8-2)

This knot is for putting loops at the end, or anywhere along a rope. It is generally easy to untie, and on average maintains 85% of the breaking strength of the rope. For putting a loop at the end of a line, simply double the rope back on itself and tie a Figure-8 in it, as shown in Fig. 8-2B. To form this knot around an object, such as an anchor, put a basic Figure-8 in the line (Fig. 8-2A), take the tail around the object, and feed it back into the Figure-8 in the opposite direction (Fig. 8-2B). If done correctly, it will look like one big Figure-8 knot. For maximum strength, the tail of the rope should be in the center of the knot, not near the outside, as shown in Fig. 8-3.

Fig. 8-2A

Fig. 8-2B

Fig. 8-3
WRONG

BOWLINE
(Fig. 8-4)

This knot is used for putting loops at the end of a line, but its primary use is attaching rope to anchors. It can be easily tied around trees and poles; yet it is strong and easy to untie. The tail of the rope must come out on the inside of the knot – if it comes out on the outside, you've tied it wrong! For safety, the tail should be secured to the Bowline's loop with an Overhand knot (Fig. 8-5). The average strength of Bowlines is 76% of the strength of the rope.

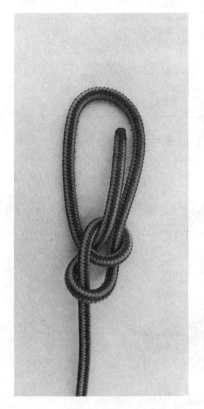 

Fig. 8-4 **Fig. 8-5**

FISHERMAN'S KNOT
(Fig. 8-6)

This is a good knot for joining two pieces of rope; particularly if they are of different diameters. It is a very compact knot, and is strong and not apt to come untied accidentally. As can be seen, this knot is simply two Overhand knots tied back-to-back, each on the other's line.

Fig. 8-6

DOUBLE FISHERMAN'S KNOT
(Fig. 8-7)

This knot is similar to the Fisherman's, except each side of the knot has one extra loop. This knot is extremely secure, and quite strong. It has 75% of the breaking strength of the rope itself. It is excellent for joining pieces of webbing and is perhaps the most secure knot for use with rope.

Fig. 8-7

SQUARE KNOT
(Fig. 8-8)

This knot is used for tying two lines together while cinching them (like in tying your shoes). On both sides of the knot, the lines must come out on the same side of the adjoining loops. If they come out on opposite sides, you have a Granny knot – which will not hold (Fig. 8-9). Rope ends, on either side of a Square knot, must always be secured with Overhand knots or Half Hitches. Never use this knot for connecting standing lines – like rappel ropes; use a more secure knot. Average strength of a Square knot is approximately 55% of the rope's strength.

RIGHT
Fig. 8-8

WRONG
Fig. 8-9

WATER KNOT
(Fig. 8-10)

Also commonly known as the Ring Bend, this knot is excellent for joining webbing; but can also be used with rope. As can be seen, it is simply an Overhand knot followed in the reverse direction with another. When you are finished, it looks like one big Overhand knot. The average strength of this knot is about 78% of the breaking strength of the line.

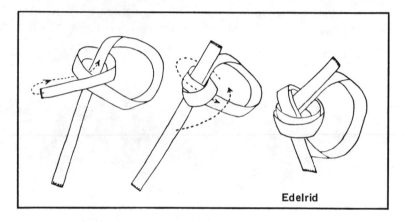

Edelrid

Fig. 8-10

Many rappel accidents are caused by knots coming untied. To reduce the chance of this happening, leave the ends of the rope, coming from the knot, as long as practical. For extra safety, tie them around the main line, using Overhand knots.

ANCHORS

An anchor is what you tie your rope onto – your rappel is no safer than the strength of your anchor. Because most rappel accidents in climbing and mountaineering result from anchor failure, it is extremely important to know how to set up safe and sturdy anchors.

The worst rappel accident on record was caused by anchor failure. Three climbers, preparing to rappel, had their ropes attached to an anchor which consisted of two bolts secured into a rock wall and connected by a chain. A sack of equipment was attached to one of the ropes and tossed over the edge. The impact force, generated as this rope became taut, was enough to break the anchor – and pull the men to their deaths.

The most used anchors are trees, and they are usually safe if large enough. Avoid using limbs when possible. Even very small trees can sometimes be used safely if tied onto at ground level. If a tree is to be used often as an anchor, it will live longer if you permanently affix a chain around it, and connect your rope to the chain. Repeated pulling of ropes from around a tree will cut the bark and kill the tree, making it unsafe for future rappels.

Fig. 9-1

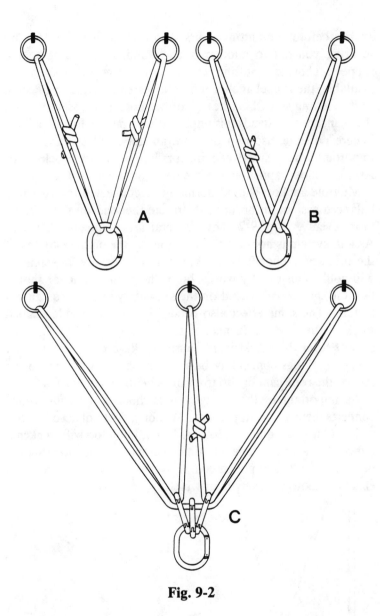

Fig. 9-2

In many rappel situations, a good safe tree will not be available, and it will be necessary to connect onto chocks, pitons, bolts, nubbins, pipes, chimneys, or etc. If possible, test the anchor before committing yourself to it. This can be done by securing yourself to another anchor, attaching a sling to the rappel anchor, and putting weight on it. If no other anchor is available, the rappel anchor might be tested by attaching a sling to it, hooking something heavy to the sling, and tossing it over. If the anchor withstands this impact, it may be safe for rappelling. Where possible, use two or more anchors. This is especially important when using weak anchors like pitons or chocks. Air Assault School rappelling is done with two anchors (Fig. 9-1).

Multiple anchors should normally be connected together in a balanced system, in order to distribute the load on the anchors more evenly. Fig. 9-2 shows several arrangements. The Air Assault system is not balanced. It goes by the philosophy that if the primary anchor fails, the secondary (normally unweighted) will hold. This usually works, but if the primary anchor should fail, the shock load placed on the secondary might also cause it to fail. The same effect also occurs, because of sudden added slack in the lines, if an anchor fails in Figs. 9-2B or C. The impact from this slack could cause breakage of slings or other anchors. Slings connecting balanced anchors should be rigged so that the rope cannot slip from the sling if an anchor fails.

As important as the strength of anchors is the sling which connects them to the rappel rope. Webbing does not absorb shock well, and it wears quickly. Rope is better, but it too will weaken if left outdoors for any length of time. Permanent anchoring should be done with heavy plated chain. Never use webbing, rope, or chain of unknown strength, or which appears worn.

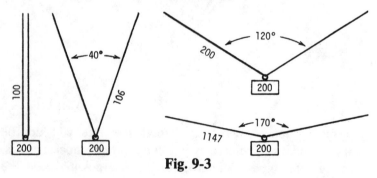

Fig. 9-3

The angle at which anchor lines run is also of great importance. As can be seen in Fig. 9-3 , when lines run parallel, each supports one half of the attached weight (load). As their angle becomes greater, the tension on each line can become much greater than the total load. At 120°, the tension on each line is doubled; at 170° it is over eleven times greater. This can be calculated by means of the formula:

$$T = \frac{L}{2 \cdot \cos \left(\frac{\Theta}{2}\right)}$$

"L" stands for total load suspended on the anchor. "L" is divided by twice the cosine of half the anchor angle Θ (theta). The result "T" is the tension on each line.

Make sure that all ropes and slings in an anchor system are capable of withstanding the stress of both static and dynamic loads. Keep anchor lines as near parallel as practical, and avoid running them over sharp or abrasive edges.

Another factor to consider is how rappel ropes are attached to the anchor. Fig. 9-4 shows a few ways. The simplest method is to loop the rope around the anchor and rappel down double-line. This makes rope retrieval easy. However, if one side of the rope should break or be cut, you would fall to the ground as if there was only one rope. It is safer to anchor both sides of the rope.

When rappelling single-line, the rope can be attached to the anchor with a Bowline or other safe knot. If speed is of great importance, a loop can be tied in the end of the rope ahead of time, and a carabiner clipped into it. When it is time to rig, the rope is simply taken around the anchor, one or more times, and the carabiner clipped onto the standing line; as shown in Fig. 9-5B. If the ultimate in strength is required, as when rappelling on thin or brittle single-line, use as large a diameter anchor as possible and put many turns around it. The bigger the anchor, and the more the turns, the less chance there will be that the rope

Fig. 9-4

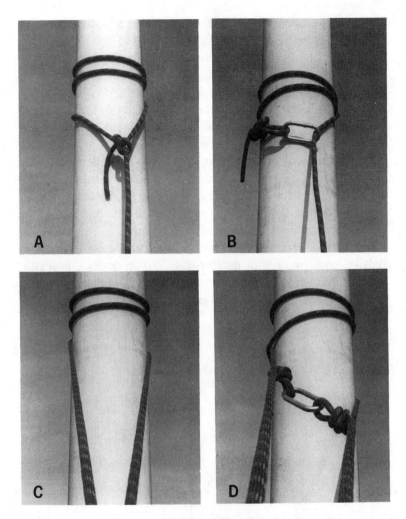

Fig. 9-5

will break. If enough turns are used, the end of the rope will have little or no stress on it, and can be tied off as desired.

If it is necessary to connect onto a smooth pole or post, you may find it difficult to keep the rope from sliding down, when placing the anchor point high. The solution to this problem is to wrap the rope "Prusik style" around the pole. If enough turns are used, the rope will stay where you put it no matter how slick the pole may be. Fig. 9-5 shows a few ways to do this.

In some situations, there is virtue to anchoring the rope with Prusik knots or other rope locking system (Fig. 9-6). The Prusiks may apply less stress to the rope than a knot, plus they can be used to form a balanced anchor system, similar to that shown in Fig. 9-2A. They also permit easy adjustment of the rappel rope's lateral position on the edge of the drop.

For safety, it is important to use a heavy Prusik sling – and preferably to use two in parallel. This way, if something breaks, there will be a backup. Although it can be totally suspended by slings, it is best if the rappel rope itself forms one leg of the system. This is both safer and more economical.

Mechanical ascenders and rope brakes can be used instead of Prusiks, sometimes with great virtue. Some devices, however, may do damage to a rope during conditions of heavy loading. Prusiks generally do no damage, and will grip the rope more securely under a wider variety of conditions.

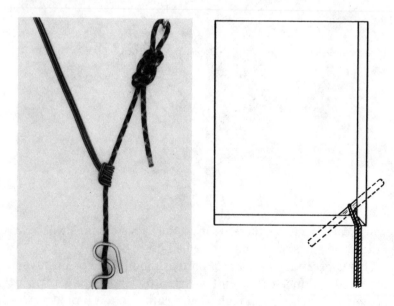

Fig. 9-6   **Fig. 9-7**

In the majority of rappel situations, there are sturdy objects nearby on which to secure the rope. In certain emergency situations, however, more exotic anchors must be used. One example is shown in Fig. 9-7. Here, the rappel rope is looped around a sturdy bar, which is then placed diagonally in the lower corner of a window. Using extreme caution, a rappeller can hold the bar in place while sliding out of the window.

Even better and safer is placing a heavy object, such as a piece of furniture, across the window and anchoring to it. However, all such techniques must be considered dangerous because of the possibility of the anchor breaking or slipping thru the window. Never use such anchors except in an emergency.

A useful way to anchor a rope, during rappel training sessions, is with a descender (Fig 9-8). Attach the descender to a sturdy anchor at the top of the drop, and rig the rappel rope thru it as if it were a belay system. The purpose of this method is so that the rappeller can be quickly and safely lowered to the ground in case of a jam or other problem.

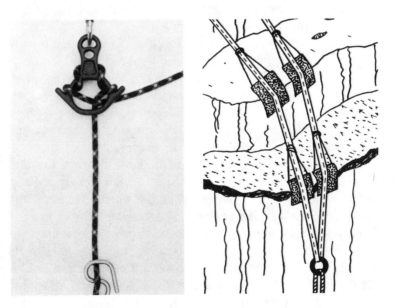

Fig. 9-8   **Fig. 9-9**

In order for the rappeller to be safely lowered, the length of the rappel rope must be twice that of the drop, if the belay is to be controlled from the top, and three times the drop height if it is to be controlled from the ground. The rope must be securely locked off during the rappel, so that it will not slip and let the rappeller fall. However, it must be locked in such a way that it can be released quickly in case of a problem. If locking is to be done by means of the descender itself, make sure that it cannot come unlocked as the rappeller swings about during descent.

During any operation of this nature, only the most competent people should be allowed to operate the system – and they should be wearing heavy leather gloves. In most situations it would be wise to run the rope thru another belay system, before it goes to the anchor descender. This will permit more control during the lowering process, and make locking and unlocking easier.

When doing multiple rappels, slings connecting the rope to the anchor are repeatedly stretched, and may get abraded on rough surfaces (Fig. 9-9). If possible, arrange things so that the sling does not contact rough surfaces. If necessary, use padding to protect the sling and the rappel rope.

Pulling a rope over a webbing sling, during retrieval, is usually easy if everything is dry, but can lead to abrasion and failure of the sling if done several times. In order to prevent such abrasion, the rope can be attached to the sling with a descending ring or carabiner. It is best if this connection point is not against a rough surface, as this can lead to excessive abrasion of the rope and sling, plus increased difficulties during retrieval.

Remember that anchor failure is the most common cause of rappel deaths. If it is necessary to use an anchor already in place, test it carefully before placing your life on it. Be especially cautious when hooking to webbing or rusted bolts. They are usually well weathered, and may be near the point of failure.

When using rocks for anchors, bear in mind that knobs as big as your head can sometimes be broken off easily. Push or kick the knob, if possible, to determine if it is sound; and make sure the anchor sling can't slide off. When using knobs or any other anchor which may not be bombproof, remember to rappel gently.

ON BELAY

Belaying is a way to provide a measure of safety if a rappeller should lose control, or his rope should break. It can be done to some extent by the rappeller (Chapter 13, "Securing a Rappel"), but it is best to have the belay done by another person with another rope.

Before getting into types of belays, there are a few rules which everyone should understand: The belayer must always wear thick leather gloves. When top belaying, a harness should be worn and it should be tied into a sturdy anchor – preferably a different one than that being used for the belay or the rappel rope. The belayer must never let go of the braking end of the belay rope, and must maintain the belay until the rappeller is safe.

The rappeller and belayer should have clear communication between themselves. Some commonly used signals are: "On Rappel," which is yelled by the rappeller after he hears an "On Belay" from the belayer; and "Off Belay," which is yelled after the rappeller signals he is "Off Rappel." In case of a fall, the rappeller should yell "Falling." When voice communication is impossible, and radio equipment isn't available, signalling can sometimes be done via jerks on the belay rope.

Belays can safely be rigged using static rope if the belayer is above the rappeller, and no long falls occur. If the belay might have to hold a hard fall, like in rock climbing, then dynamic rope should be used. A good rule is this: If the belay anchor point is above the climber or rappeller, static rope can perhaps be used – if the anchor is below the climber, use dynamic rope.

The simplest and least valuable belay is a Bottom belay, whereby the belayer stands on the ground below the rappeller (Fig. 10-1). By simply pulling down hard on the rappel rope, the belayer can normally stop the rappeller's descent – unless the rope breaks or comes untied. Bottom belays will work with most descenders (a notable exception being Pressure Plates), but they become less effective as rope lengths become greater. On rappels of several hundred feet or more, rope stretch can make a bottom belay totally ineffective – even when the belayer's full weight is on the rope. Bottom belays must never be used during body rappels, for they would cause the rappeller to fall.

Fig. 10-1

Fig. 10-2

The best belay is a separate rope from above, which is attached to the rappeller's seat harness. The rope is payed out as the rappeller descends, allowing a little slack. The belay rope should be run thru a rope protector or over a padded edge, so that it won't be cut. It should be hooked to a different anchor than the one holding the rappel rope; and should be rigged so that the belayer can lock off, if necessary, and go for help.

On top, the belayer must not simply hold the rope in his hands, or wrap it around his body (Fig. 10-2). Hard falls can seldom be held by these methods, and can easily result in the belayer being rope-burned and the rappeller being killed. Such "body belays" can be very dangerous.

To provide friction and prevent the belayer from holding the entire force of a fall, the rope must be run thru or around a friction producing device. A good belay setup will enable the belayer to easily stop a falling rappeller with one hand. Perhaps the simplest way to do this is to wrap the belay rope around a sturdy tree or post, one or more times (Fig. 10-3). This doesn't provide the smoothest belay, but it is normally safe.

Fig. 10-3

Most descenders can be used to belay. They are used exactly as in rappelling, except that they are attached to a rigid anchor instead of a rappel harness. The cover photo of this chapter shows a girl operating a belay which uses a Sticht Plate to provide friction. It is a very easy system to manipulate, yet provides tremendous amounts of friction the moment it is needed. Note: Sticht Plates work best when used with kernmantel rope. A Half Ring Bend, which is the U.I.A.A. method for belaying, is also an excellent way to obtain friction. For more information on this system, see "Half Ring Bend" in Chapter 6.

Top belays can pose certain dangers, such as causing rocks to fall. If the belay rope dislodges rocks, the rappeller can be injured or killed. In areas of heavy rockfall, such as in certain caves, it is often wise to not use top belays. Besides injuring the rappeller, falling rocks can also injure people on the ground, and even cut the rappel rope. People on the ground, in dangerous areas, should always wear helmets and stand clear of the rockfall line. Because the rappel rope can also dislodge rocks, bottom belayers should be warned to stand clear of fall lines.

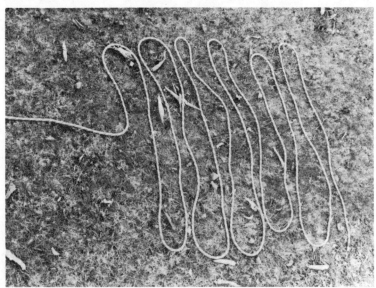

Fig. 10-4

Another problem has to do with rope twist. Because the rappeller usually spins in the air on long, free rappels, the belay and rappel ropes can become tangled. This occurs most often when rappelling on laid rope, but can happen with kernmantel. One remedy is to rig the belay a substantial distance from the rappel anchor, so that the ropes do not run in parallel (Fig. 10-5). Although this can prevent the rappeller from spinning, it can prove hazardous if the rappel rope breaks. In such an event, the rappeller would pendulum sideways, possibly hitting rocks or cutting the belay rope.

With any top belay, the rope needs to be managed in such a way that it will not kink or jam up the belay system. It must pay out of its coil smoothly. The Fake/Bag or the Monkey Chain are excellent ways of handling the situation (see Chapter 3, "Rope Handling"). One of the best methods is faking the rope across the ground behind the belayer (Fig. 10-4). When laid out in this fashion, the rope normally pulls without kinking.

If an extra rope isn't available for a belay, the tail of the rappel rope can sometimes be used to provide safety for getting over the edge (Fig. 10-5). Simply pull up the rope, from below the rappeller, and run a loop of it thru a belay device. Connect the end of the rope to the rappeller's harness. This belay is managed like any other, and can be kept in operation for about 1/3 the length of the rope. Once the rappeller is safely over, the belay is disconnected and the rope carefully let down. Never drop it! If the rope is heavy, dropping it might generate enough force to damage the descender or pull out an anchor.

The purpose of any belay is to reduce the chance of injury or death, in the event of a fall. In order to accomplish this end, it is important that falls be arrested quickly, in order to keep the rappeller from hitting anything. However, it is also important that the stop not be too quick, lest the rappeller be injured or anchors be broken. When belaying with static rope, or any type of low stretch line, use a belay device which will not suddenly lock and put undue strain on the system. Belays should be managed in such a way that falls are arrested gently, with little pain to the belayer or rappeller.

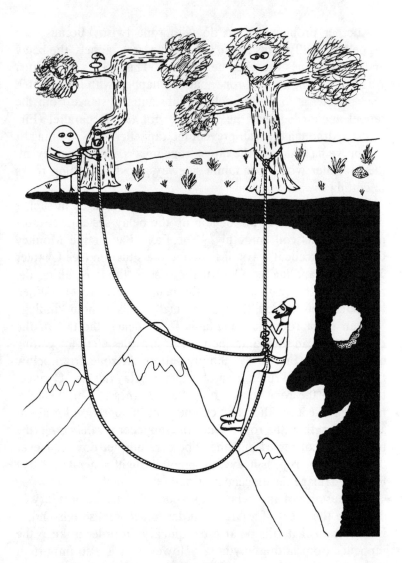

Fig. 10-5

GETTING OVER

The most hazardous and frightening part of the average rappel is getting over the edge. Once you're over, it's usually smooth sailing the rest of the way down. Some edges are very easy and safe, but others are so tricky that they make most experienced rappellers have second thoughts about the sanity of going over.

Edge difficulty depends on many factors, such as shape, sharpness, and brittleness of the edge material. The edge can also be made treacherous by things like moss or ice. Difficulties are magnified by the height of the drop. The higher it is, the worse it seems, and the more the fear is felt. It is this fear which can lead to mistakes – and to death.

A big, if not the biggest, factor in edge safety is the angle of the section of rope between the anchor and edge (Fig. 11-1). Imagine that you are going to rappel down a cliff. You are standing next to the edge and, by some miracle, there's an anchor point above your head. You tie onto it and rig for rappel. Since the anchor is directly above the edge, the rope runs vertically from the anchor to the edge. As you start over, you will note that there is no way you can lose your balance. The edge could be covered with moss, or ice for that matter, and it wouldn't matter a bit since your feet aren't supporting any of your weight. This is a perfect rappel situation.

Although you will seldom, if ever, find yourself in such a situation, you should always strive to make your rope run as near vertical as possible. Even if it means climbing a tree, try to get your anchor point as near vertical above the edge as your rope length, anchor strength, and situation safety will allow. The more horizontally the rope runs, from anchor to edge, the greater your chances of an edge accident.

When we speak of edge accidents, we mean all the things that can happen when you lose your footing while going over (anchor and equipment accidents are not included in this definition). The most common edge accidents occur when the rappeller loses his balance and falls sideways. Most other accidents occur when one's feet slip. Most of the accidents result in abrasions and bruises from hitting the rock. Sometimes they end in death if the rope gets cut, or the rappeller loses his grip on the rope.

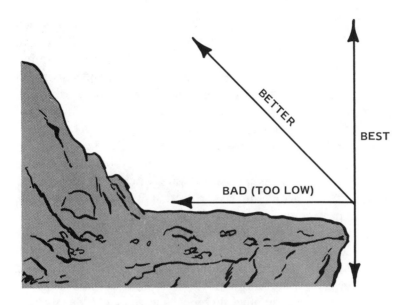

BETTER

BEST

BAD (TOO LOW)

Fig. 11-1

Most accidents could be prevented if the rappeller would put the anchor high and keep his feet spread apart for balance. Foot slippage, usually occurring on a slippery surface, might be minimized by wearing the proper footgear for the situation. Edge accidents are good reasons for safety helmets and belays.

The problem of losing grip on the rope is a very real one. If they stumble while going over, many rappellers have a tendency to let go of the braking end of the rope and grab hold above the descender. If there is little friction in the system to start with, the rappeller finds himself supporting almost his entire weight with his hands – sometimes with disastrous results. A thin rappel rope is awfully hard to grip. Train yourself to grip the braking end of the rope harder if a fall occurs – never let go of it. You can also help minimize this danger by using a dead-man brake (see Chapter 13, "Securing a Rappel").

The problem of rope cuts are a subject unto themselves. Most ropes that break do so because they have been cut. Sometimes a rope can be cut by just sliding a few inches across a blunt edge;

like what happens when one falls sideways while trying to get over. Because of this great delicacy, the rope must be handled very carefully. How do you protect the rope? First, be careful! Then, if possible, make sure the edge is rounded and smooth. If the edge isn't to your liking, look around and you'll probably find a better place a few feet away. If you can't move and you have to go over – and if and only if it is an emergency – take a hammer or rock and round off the edge until it appears safe.

Next, before you go over, put something soft between the edge and your rope. You can use any kind of clothing or fabric or leather. Some people slip a short section of garden hose over the rope to protect it. This is great if the bulk and weight of the hose is tolerable. Others take soft plastic tubing and split it down the side so that it can be slipped on and off with ease. This works, but it has the same shortcomings as a cloth pad – that is, if you should fall sideways or somehow move the rope, it will move off the cloth or it will get out of a split tube – and you will suddenly have no protection.

Perhaps the most satisfactory method of protecting the rope is to slip a two or three foot (1m) section of one-inch tubular nylon webbing over it (see Chapter 7, "Rope Protectors"). As you go over the edge, just slide the protector down with your balance hand until it is in a position which will protect the rope from the edge (Figs. 11-6 – 11-9). Since the webbing is flexible, it can be left on the rope at all times. Since it never has tension on it, it is very difficult to cut and therefore offers great protection. Each time a rope touches rock, fibers are damaged; by using a rope protector, you will find your rope lasts much longer. Be sure and carry two protectors for double-line use.

Another of the many factors which affect safety is the shape of the edge. There are three basic edge shapes with which the rappeller has to contend, whether going down a cliff or a building. The easiest type is a slope or incline. (A slope is a safe place to learn rappelling). Actually, going down a steep slope isn't always rappelling; but the same equipment and techniques are used – and slopes have a way of turning into rappels, particularly on dark nights.

The next most difficult edge occurs on a wall which is more or less vertical for several feet below the top. To safely go down, you should place the anchor as high as possible, test it, hook up, and carefully back up to the edge. Plant your feet squarely on the edge and spread them apart – so that you won't lose balance and fall sideways – and bend your knees slightly. Without moving your feet, slowly lean back. Keep leaning back until you are standing almost perpendicular to the wall (Fig. 11-2). Gently, one step at a time, start backing down the wall. Your weight should be pressing your feet against the wall in the same manner as if you were standing on a floor.

Fig. 11-2

If, as you back down the wall, most of your weight is supported by your toes, your feet could slip downward – lean back more. If your weight is on your heels, you are in danger of doing a backflip, so move your feet down a bit. If the wall is vertical, but the edge is very rounded, there may be no need to carefully plant your feet and lean back. Just carefully walk backwards over the edge while maintaining balance as described.

The most dangerous edge is one which is very sharp and goes back under itself. This is known as an overhang or knife-edge. You can't walk down this one, as you can a wall, because below the edge there are no usable footholds. Although sharp edges are the most dangerous, even a very rounded overhang can be treacherous, and requires special techniques for going over safely. This is the type of edge you sometimes see the "experts" jump off of. That's one reason "experts" sometimes get killed rappelling!

If you must go off an overhang, and the anchor-to-edge part of the rope is running at a low angle, try to raise the height of the anchor. If you can't, then prepare to sit down near the edge and slide over sideways (Figs. 11-3 and 11-4). This isn't very classy and it usually is painful; but at least it minimizes the chance of a cut rope, which might occur if you try going over in the normal standing position. The problem with crawling over is that the descender tends to get hung on the edge. When this happens, you can't rappel down; and if you don't have enough strength to pull yourself back up, you are stuck. When sliding over in this fashion, getting hung is the rule rather than the exception.

This problem can occur with any descender; but occurs most often with descenders which project out from the body, such as Racks and Snakes. It also happens frequently with Figure-8's; especially those with ears or horns.

If you get hung on an edge, remember the Prusik loops you carry in your pocket for self-rescue (see Chapter 12, "Prusik Method" and Chapter 13, "Getting Unjammed"). By connecting onto the rappel rope, above the descender, you can provide yourself with a foot loop with which to take the weight off the descender. Once the descender is moved off the edge, you can either climb back up and start over, or continue the descent.

Fig. 11-3

Fig. 11-4

This method can also be used to safely get over the edge, and it sure beats trying to crawl over. Simply attach a Prusik or other ascender to the rope, before you start over, and position it so that the foot loop is a convenient distance below the edge. An etrier is excellent for this since it can be used as a ladder and a Prusik (Fig. 11-5). To go over, simply hold onto the rope, carefully put your foot in the loop, and use it like a ladder to climb over the edge. Once over, and hanging safely from your descender, undo the Prusik and continue the rappel. Never climb down over the edge without first having your descender properly rigged and attached to your harness and the rope.

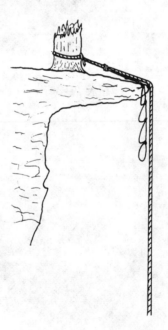

Fig. 11-5

The next three paragraphs describe a unique way to rappel off an overhang in a normal upright position. This method is presented, not as a recommendation, but simply as an alternative when other methods of getting over prove impractical or dangerous.

Bear in mind that if the anchor-to-edge part of the rope is not running at a high angle, this technique becomes very hazardous. If the rope angle is too low, you will reach a certain point when leaning back where you no longer have control. This occurs as you become approximately horizontal. When this point is reached, you will lose control and swing back under the overhang. If the edge is sharp, the rope may get cut. You may also hit rock outcrops below the overhang and be injured. You must practice this technique on a low, safe overhang (with belay and safety helmet) until you learn which rope angles will give you control and which won't. Remember that many factors affect how well this technique works, including rope stretch and system friction, and these factors will change in different situations. This is one of those rappels where low stretch rope is almost mandatory. Under no circumstances try this technique while wearing a type of harness which you can fall out of – and always use protectors on the rope.

Here is the technique: Simply plant your feet on the edge and spread them apart for balance. If the overhang is really sharp, place your feet so that your heels hook onto the edge, and lean back (Fig. 11-6). Keep leaning until you are entirely upside-down and standing on the bottom side of the overhang (Figs. 11-7 and 11-8). Do this slowly and smoothly so that the rope will not slam into the edge. Once you're upside-down, drop down a few feet, right yourself and continue the rappel (Fig. 11-9).

If the overhang is quite rounded, you may have to walk backwards over the edge until you are upside-down (Figs. 11-10 and 11-11). Either way, this technique solves the problem of what to do when there are no footholds below the edge, and can make this type edge a pure joy to go over. Just don't forget its dangers! Read the "Warning" again, two paragraphs back.

One of several exceptions to this method of handling overhangs is when rappelling out of helicopters. Because stable hovering is not easy in some wind conditions, and because of the danger of hitting the copter's landing gear, it is sometimes advisable for the rappeller to jump from the copter. When the "jumping" method is deemed proper, it should be done in this fashion: Connect your rappel rope to the anchor points in the copter's ceiling, and then

Fig. 11-6

Fig. 11-7

Fig. 11-8

Fig. 11-9

Fig. 11-10

Fig. 11-11

carefully check to make sure the metal plates in the doorway are smooth and rounded. If things look safe, and speed is necessary, place your feet on the edge and leap backwards with considerable force. There should be lots of slack in the rope. Keep your balance arm directly in front of your face and be prepared for it to get smashed into the doorstep on your way down. Also, before you hop out, make sure your rope reaches the ground. In the event the copter pulls up too quickly, be prepared to instantly stop the rappel, lest you find yourself in midair – with no rope!

If the helicopter has landing skids, it may prove safer to climb down onto a skid to begin the rappel, instead of jumping from the doorway. Just remember that the copter can lunge suddenly and throw you off the skid. All helicopter rappels, particularly those from small copters, should be done as smoothly as possible. Long drops and sudden stops can throw a small copter out of control. Very elastic rope is sometimes recommended to reduce these shock loads, but in practice this can cause accidents. As the rappeller bounces up and down on the end of this elastic line, the helicopter pilot loses more and more control. One large jolt is usually not hard to handle, but a series of smaller tugs can bring disaster. All rappels from small helicopters should be done with low stretch static rope.

The techniques described for getting over the various edge types can be applied in many cases just as described. But you'll find that the average edge is sort of a combination of all three types, and therefore the technique must be worked out as you go. Use this book as a guideline, but above all, use your head.

Danger in any edge situation is minimized by having a quality harness which fits properly. As detailed in Chapter 5, a harness should provide a snug fit without being uncomfortably tight. The point where the descender attaches to the harness should be near your body's center of gravity. This makes for better balance and increased safety when going over tricky edges such as overhangs. No matter what the quality or cost, you should switch to another type harness if the one you are now using does not allow you to hang in proper balance. This is especially important if the harness makes you feel extremely bottom-heavy.

One of the main dangers to be faced when going over is caused by rope stretch. If you are using a rope with lots of stretch, it can cause you to lose your balance. Here's what happens: As you start leaning back, there isn't much weight on the rope. The more you lean back, the more weight will be put on the rope and the more it will stretch. If you're using a low stretch static rope, the amount of stretch will be so small that it probably won't be noticed. If you happen to be on a piece of high stretch lay rope, you may find that the rope suddenly acts like a rubber band. When this happens, you may suddenly be let backwards more than you intended, and possibly lose your balance. This is one reason why rappel ropes should preferably be of the low stretch variety.

The problem of stretch is compounded by long anchor-to-edge rope lengths. Whatever type rope you use, try to get the anchor point close to the edge; but not so close that it makes going over unhandy or hazardous. Rappelling double-line and using large diameter rope will also help reduce the amount of stretch.

The height of the drop is a factor which can cause trouble. When a cliff is high, the weight of the rope hanging over the edge can cause great problems. On a long drop, one would have to fight the entire weight of the rope while going over; weight which can be greater than that of the rappeller. This rope pull causes problems because it tends to force you down on your knees at the edge, making it difficult to maintain the proper stance.

The rope pull also increases the friction in the descender, making it difficult, if not impossible, to descend (see Chapter 6). The solution to this problem is to use a descender which allows you to make large changes in friction. Attach it to the rope in such a way as to provide minimum required friction, then as you move down the rope and have less rope weight below you, merely add more friction to the system as needed. This rope weight problem is why most rappellers choose Racks when doing extremely long rappels. Racks make friction adjustment simple. The problem can be helped to some degree by tying the tail of the rope to your harness carabiner, before you go over. This cuts the rope pull on the descender by half and effectively increases your weight. Both factors make moving the descender easier.

Fig. 11-12

A simple way to eliminate rope pull entirely is to have someone at the top pull several feet of the rope back up over the edge, before you start the rappel (Fig. 11-12). The short unweighted loop formed below you will allow getting over with ease. When you're safely over, your companion can carefully let the rope down. You then adjust the descender's friction to compensate for the added rope weight, and continue down.

As simple as this system sounds, it can have problems. The first can occur if the rope is dropped. As the rope gets taut at the end of its fall, it could generate enough force to damage the descender or pull out the anchor. **NEVER** drop the rope! Both the rappeller and the person on top should cooperate in such a way as to let the rope down slowly. A second problem occurs when using fixed-friction descenders such as Figure-8's and etc. These devices would allow getting over the edge with no problem; but once the rope was let down and its full weight applied to the system, you might find it difficult, if not impossible, to descend. One solution would be to actually pull yourself down the rope. This is a lot harder than it sounds!

But, you might ask, why not use a low friction device to start with? This sounds good, but what do you do when you get near the end of the rope and there's no more heavy pull to compensate for the low friction in the descender? You could try to wrap the rope around parts of your body for more friction – if you have time. Long drops should only be done with variable-friction descenders.

Another problem to be considered is edge strength. Edges do sometimes break – particularly those thin overhangs. Before commencing a rappel, examine the edge and try to test it for strength. Do this is such a way that you won't go tumbling down with the pieces if it does break! By all means, don't equate thickness with strength. Just because an edge is several feet thick doesn't mean that it will hold your weight. Test it!

Don't forget the other great danger of going over. Clothing and hair can get caught in the descender. If you have medium to long hair, it can, and probably will get caught. This can happen while going over or anywhere down the line, and it only happens when you're not thinking about it. You can prevent this painful experience by being careful. This is one of the reasons for carrying prusiks – and also the main reason for carrying a knife or scissors. If hair gets too tangled in a descender, it may be necessary to cut it free. Just be careful and don't cut the rope. Whenever possible, use prusiks to effect the rescue.

Most jams occur, not from hair, but from clothing getting pulled into the descender. This usually poses few problems, and getting free is normally easy. However, there is sometimes pain; particularly if flesh is pulled along with the cloth. Folks of the feminine gender should beware of getting their breasts close to descenders! Also, be especially careful of clothing or slings draped about the neck – if such things get pulled into the descender, you may die.

GOING UP

Rappelling is the art of going down, but there are times when you may need to go back up the rope. In many caving and rescue operations, the only way out is up. Also, if something jams during a rappel, you may very well have to ascend – unless you like hanging there.

Because this book is about rappelling, no attempt has been made to cover the entire subject of ascending. The methods and equipment discussed are those deemed most practical for rappellers. Practice these methods well – before you have to use them. Also, just as you should never rappel on a worn rope, don't ascend one either. They can break – or in the case of old kernmantels, the mantel can break, allowing the climber to slide quickly to the ground.

Ascending can be done by the old hand-over-hand method, but this is only practical for a few feet. Never use this method to climb any distance. Many folks have died trying! There are many devices available which make ascending simple and safe. They lock onto the rope when weight is applied, but can be easily moved upward when the weight is removed. By using these ascenders in various ways, any length of rope can be safely climbed.

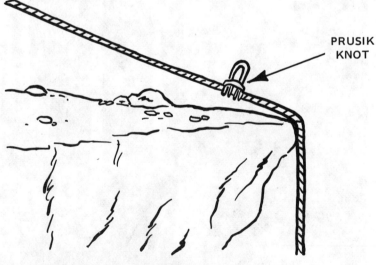

PRUSIK KNOT

Fig. 12-1

This chapter details some of the more commonly used ascending equipment and methods. They vary in speed, efficiency, and cost, but all are normally safe – if used correctly. Besides carrying enough ascenders to climb a rope, it is also helpful to have one extra. Fig. 12-1 shows how this ascender can be attached to the rope, above the edge, to make getting up easier and safer.

One of the ascending methods described is the Texas Inchworm, and every rappeller should be skilled at it. It requires the least equipment, and in a bad situation may enable you to climb out, using only your descender and a shoelace.

Before learning ascending methods, you should have a good knowledge of ascending devices. The ascenders you use should be chosen for the job to be done. As with all equipment, you should test their strength and check their function before use.

ASCENDERS

KNOTS

One of the oldest, safest, and cheapest ways of going up a rope is with sliding hitches. There are many types of these knots, but the most used are the Prusik (Fig. 12-2), the Heddon or Cross Prusik (Fig. 12-3), and various Single-line Hitches (Fig. 12-4). These hitches can be used on single or double rope, and when weight is applied to their slings they tighten up on the rope. They can be slid upward by hand when the weight is removed. These Hitches can normally be used to climb a rope which has tubular nylon rope protectors placed on it.

Hitches should be tied with strong, flexible, low stretch cord which is smaller in diameter than the rope. 1/4 inch (6mm) is a good choice for use on 7/16 inch (11mm) rappel rope. Larger cord can be used by putting more turns in the hitch. More turns are also needed if the rope is icy or muddy. Polyester cord is best for hitches, although nylon works well. Kevlar can be used, but if a hitch jams, cutting it free could be difficult or dangerous.

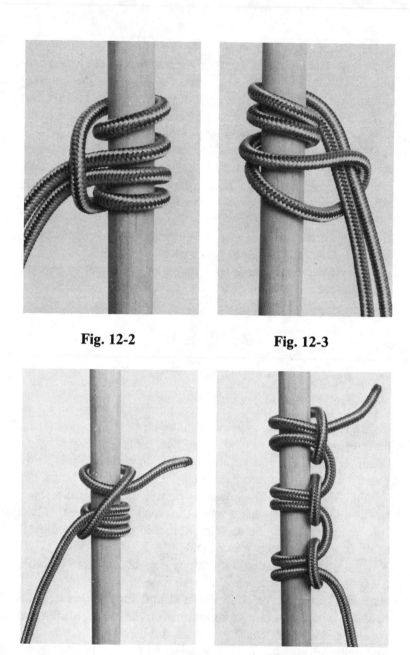

Fig. 12-2   Fig. 12-3

Fig. 12-4

Because sling cord is small in diameter, it is extremely important that it be of good quality and in good shape. New 6mm nylon cord will hold from 1,200 to 1,650 lbs. (544 to 748 kg.); 7mm will support about 2,000 lbs. (900 kg.). However, these thin cords age faster than full diameter rope, so check them often and replace them at the slightest sign of wear.

The hitches shown in Fig. 12-4 are tied with a single line instead of a loop. They will grip when no other ascending device will. Each hitch progressively increases the friction on the rope, and can be added until sufficient friction is obtained. The problem with this knot, or any other with too many turns, is that it may be difficult to unlock it and move it up the rope when the weight is removed. The only remedy is to remove some turns.

Besides sling cord, ascending knots can also be tied with nylon webbing; sometimes to much advantage. In an emergency, knots can be constructed with shoelaces, belts, hair, strips of cloth torn from clothing, and etc. If one lace is not strong enough, two or more can be placed in parallel.

Ascending knots have many advantages, but their forte becomes obvious when you have to climb up a rope with protectors on it. When going across tubular nylon protectors, add an extra turn or two to each ascending knot. Generally they will grip tightly enough so that the protector will not slip down the rope. Practice this trick near the ground, until you get the feel of it. If your knot will grip thru the protector here, it more than likely will do the same when you ascend.

MECHANICAL ASCENDERS

Mechanical ascenders, like knots, grip the rope when weight is applied to them. They release automatically when the weight is removed and they are pushed upward. They usually operate much easier than knots. Most mechanical ascenders work on the cam principal. Fig. 12-5 shows how the cam grips the rope when weight is applied to the ascender.

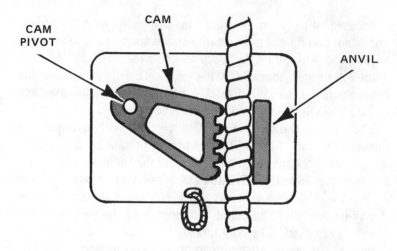

Fig. 12-5

The cam faces can be smooth, but most have ridges or conical teeth. Cams with teeth seem to cause more damage to the rope than those with smoother faces. Also, toothed cams will not grip tightly if the teeth get clogged with ice, mud, or rope fuzz.

Ascender bodies are made by casting, machining, or forming. All are strong, but cast ascenders sometimes break if allowed to fall onto a hard surface. Most ascenders will support at least 1,000 lbs. (454 kg.), although some, particularly those with toothed cams, may start cutting the rope at these weights.

The main disadvantage to mechanical ascenders becomes obvious when you have to climb a rope with protectors on it. Because most ascenders have little clearance beyond the rope's diameter, climbing over a rope protector may be impossible. Even if there is clearance, the attack angle of the cam may be thrown off to such an extent that the ascender will not grip tightly. Make sure you know what you are doing before trying to climb over a protector – lest you find yourself sliding rapidly to the ground. Having an extra ascender, to place above the protector, is perhaps the best and safest way to get around it.

The following list describes a few of the mechanical ascenders currently available:

CMI 6005

The CMI ascender, by Colorado Mountain Industries, is one of the more popular devices for ascending a rope, and can be used on ropes from 7mm to 16mm in diameter. It is machined from 6005-T6 aluminum and anodized, and weighs 8 ounces (226 g.). The cam in this ascender is made of stainless, which is then chrome plated. The cam spring and axle are also stainless.

As can be seen in Fig. 12-6, this ascender comes in right-hand and left-hand models. It can be attached to the rope, with one hand, by simply pushing the cam release lever upward and clipping the ascender to the rope (Fig. 12-7). The lever can be operated while wearing heavy gloves.

In normal use, foot slings are attached to the CMI ascender by means of the holes in the bottom of the frame. One method for doing this is shown in Fig. 12-7. The hole at the top of the ascender is for attaching slings during hoisting operations. The top hole also enables the ascender to be secured to a person's body for doing advanced ascending technique.

Fig. 12-6

When using the CMI, or any ascender of this style, be careful that the sling does not get abraded by the standing line during ascent. If necessary, move the sling to another location on the ascender body, so that it will not contact the rope.

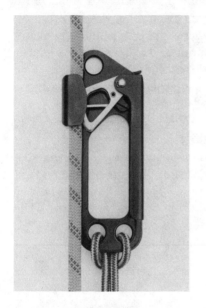

Fig. 12-7   Fig. 12-8

Shunt

Made by Petzl, this device was not designed to be used for ascending. It is used for securing rappels (see Chapter 13, "Securing A Rappel"). Because of the way it functions, however, it can be classified as an ascender, even though it is not recommended for that purpose.

The Shunt can be used on one or two ropes (of equal size) from 5 to 12mm in diameter (Fig. 12-8). It is attached to the rope, as shown, by merely rotating the cam forward in the frame, and slipping the rope(s) behind guides in the backside. Slings are attached to the cam instead of the body.

Shunts weigh 5.6 ounces (160 g.), and have bodies of aluminum sheet metal. The cams are also of aluminum, and the axles and springs are stainless. Except for the sling being attached to the cam, Shunts function like most other ascenders. They are especially useful for securing double-line rappels, and can usually be released when loaded, to expedite continuing the descent. Under heavy loading, Shunts will normally slide down the rope without causing damage. This usually occurs at about 660 lbs. (300 kg.), but varies with rope size and condition.

Gibbs

Fig. 12-9

Gibbs (Fig. 12-9) weigh 7 ounces (200 g.) and are designed for use on single rope of up to 1/2 inch (l3mm) diameter. The body is made of aluminum or stainless sheet metal, and the aluminum ridge faced cam is held in place with a quick-release pin. The cam must be removed for attachment to the rope and this requires both hands. Gibbs are available with an external spring which lightly forces the cam against the rope, thereby preventing the ascender from slipping down the rope when weight is removed. The sling cord is attached to the cam, making this ascender valuable for securing single-line rappels. Its forte, however, is climbing via the Gibbs Method. This method, described later, is perhaps the fastest way to ascend a rope.

METHODS

TEXAS INCHWORM

This method of climbing can be done with two ascenders of most any type. As can be seen in Fig. 12-10, an upper ascender is connected to the rappel harness, and lower one to the foot sling. To ascend, merely push the foot ascender upward as far as possible, stand up in the foot sling, and move the harness ascender upward (Fig. 12-11); then sit back down and repeat the procedure. With moderate effort, you can move up the rope like an inchworm. This method is nice because it allows you to sit and rest whenever you feel the need.

There are a couple of variations of the Inchworm which every rappeller should know. Fig. 12-12 shows an upper ascender and foot sling. The ascender knot can be a Prusik, or if cord length does not permit, a Single-line Hitch can be used. The other end can be tied into a loop for one or both feet; or, if the line is a shoelace, simply tie the end into a couple of eyelets on the shoe. The harness ascender is the descender you are rappelling on. Yes, you may be able to use it to climb back up the rope! Many descender types readily lend themselves to this purpose.

Fig. 12-10

Fig. 12-11

Fig. 12-12

Fig. 12-13

Fig. 12-14

Fig. 12-15

After you stand up in the foot loop, merely pull the braking end of the rappel rope back thru the descender, raising it up the rope as far as possible (Fig. 12-13). Apply tension to line (as if you were stopping a rappel), and sit down. Once again push the ascending knot upward and repeat the climbing process.

Any descender can be used for ascending if the rope can easily be pulled thru it in reverse – when tension is removed. STOPs are the easiest descenders to climb with because of their locking mechanism; Figure-8's are also desirable because they can be easily locked when resting is required. Sticht Plates do very well because they move up easily and can be held with only finger tension on the braking line.

This variation of the Inchworm is the best way to unjam Figure-8's and other descenders. It is also nice, because if the foot sling breaks, you won't fall; you will be left hanging by the descender. That's why shoelaces, strips of cloth, and even long hair can be safely used to make the sling. This technique requires the use of a locking harness carabiner or two carabiners with the gates reversed. Ordinary carabiners can easily come unhooked during the climbing process.

Another Inchworm variation is shown in Fig. 12-14. Only one ascender is required, and it is attached to the harness via a strong sling. Don't use a shoelace for this – if it breaks you die!

While hanging from this sling, raise one of your feet and loop the rappel rope under it. Bring the tail of the rope over your head and hold it against the standing line. Using both arms, stand up in this foot loop (Fig. 12-15). After you stand, move the ascender up and sit back down. Repeat the process to climb. For long hauls, it is helpful to have both feet in the loop; or perhaps to wrap the rope about the feet one or more times.

GIBBS METHOD

Also called the Rope Walker, this technique was developed for use with the Gibbs ascender, and is the fastest way to ascend a rope. The rope is climbed like a ladder, and can be climbed at more than 150 feet per minute. Also, because the ascenders are attached low on the body, this method makes getting around overhangs a simple and easy procedure.

As seen in Fig. 12-16, one Gibbs is attached at the ankle, and the other to the opposite knee. The ankle unit is tied rigidly, with two slings, so that there is little slack. The sling which goes under the foot supports all the load, while the ankle sling keeps it and the Gibbs in place. The unit at the knee is attached with three slings. One goes to the foot, one around the knee, and the other is hooked to the belt or rappel harness. This latter sling keeps the Gibbs in place when the knee is moved upward. For comfort, the slings should be made of wide webbing.

When starting out, there is little rope weight below you, and the foot Gibbs may not slide upward easily. Have someone hold the rope until you get started; or take a loop of rope under your foot to provide tension (Fig. 12-17), and pull up on it as you raise your foot. Let go of it after you've gained some altitude.

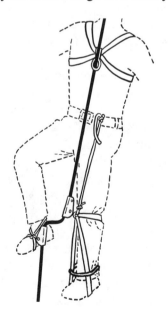

Fig. 12-16

Fig. 12-17

For safety, always wear a chest harness and hook it to the rope with a carabiner or Gibbs. This will keep you from inverting if your hands lose their grip, and will make climbing easier. When resting is desired, a Gibbs can be hooked to your seat harness.

PRUSIK METHOD

This is the classic method of ascending and can be done with Prusik knots or most any other ascender. Fig. 12-18 shows the basic method. Ascenders are connected to each foot by strong slings, which extend to just under the shoulders. A third ascender, for balance and safety, is attached above the other two. It is connected to a chest harness with a sling slightly less than the length of your arm's reach.

To climb, the ascenders are moved upward in a middle-bottom-top sequence. Normally, only the feet ascenders have to support weight. The chest ascender is for safety, although it does allow you to lean back for a rest. The long slings required for the Prusik method should be of low stretch cord or webbing, and obviously should be very strong.

Fig. 12-19 shows a variation of the Prusik method, which uses only two ascenders. It is basically the same, except that the slings to the feet are longer and are run thru a locking carabiner attached to the chest harness. It is, however, simpler and faster than the classic way. Safety slings (not shown here) should be run from each ascender to a seat harness. A Gibbs or Shunt can be connected to the seat harness to permit resting. Whatever you do, don't try climbing any distance without safety slings attached to a seat harness. Folks have died trying!

Fig. 12-20 shows one way by which slings may be attached to the feet. For greatest comfort, slings should apply equal torque to both sides of the foot. For safety, they should be secured to the legs with sturdy leg bands. This way, if your hands should slip, or the safety system fail, you will be left hanging from the foot slings. It isn't pleasant, but it beats falling!

Fig. 12-18

Fig. 12-19

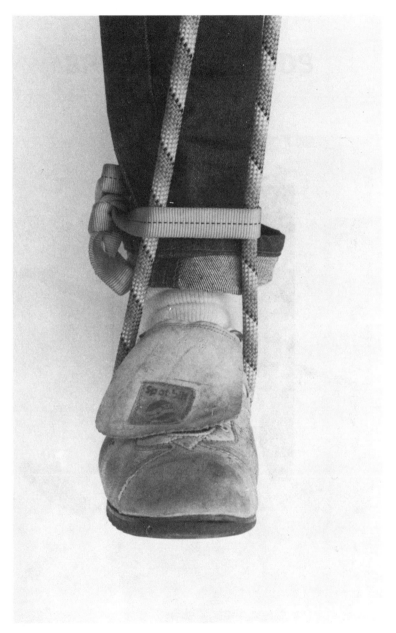

Fig. 12-20

EXTRA FRICTION

Perhaps the simplest way to get extra friction for body and descender rappels is to wrap the rope around your braking arm, as shown in Fig. 13-1. Gloves and long sleeves are a must when doing this; as is rappelling slowly and carefully. Failure to go slow could result in a well-burned arm – or one that is pulled out of socket. Extra friction can be had by running the rope across the hip or back, as shown in Fig. 13-2. This works, but can be dangerous if the rope runs across the rappel harness. On a long, fast rappel, the harness could be cut by the rope friction. Extra friction can also be obtained by wrapping your leg around the braking end of the rope, rather like that shown in Fig. 4-5.

Friction in Figure-8 type devices can be increased by "double wrapping" the rope. A loop of rope is taken thru the eye, around the tongue, back thru the eye, and looped over the tongue (Fig. 13-3). Extra friction can also be had with any descender by running the braking end of the rope thru the harness carabiner. This provides an extra margin of safety in case the descender should break or somehow become detached from the rope. Always provide yourself with enough friction. Normally, you should be able to rappel with only two fingers on the rope.

Fig. 13-1

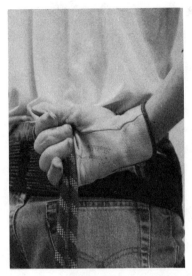

Fig. 13-2

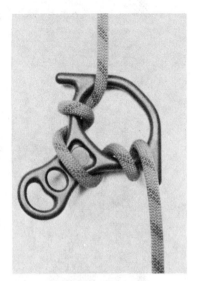

Fig. 13-3

LOCKING A RAPPEL

Locking a rappel means "locking" yourself in place on the rope, so that your hands are free for other tasks. The simplest method is for the belayer to hold the rappeller in place, but this is not always practical or possible.

As described in Chapter 6, certain descenders have inherent characteristics which enable the rappeller to quickly and safely lock in. Some descenders can't be locked, so other methods must be employed. Fig. 13-4 illustrates a variation of the Slip knot, called the Quick Lock knot, which can be used to lock most rappels. To tie this knot, push a loop of rope (from below your braking hand) thru a hole in some of the gear, and give it one full twist (Fig. 13-5). This loop can be pushed thru a large hole in the descender, or thru the harness carabiner. Next, take another loop of rope and stick it thru the first loop (Fig. 13-6). For extra safety, put a third loop thru the second one (Fig. 13-7).

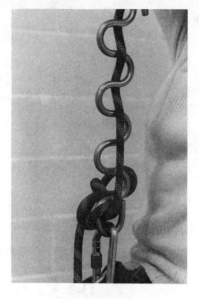

Fig. 13-4

Fig. 13-5

Fig. 13-6

Fig. 13-7

Slowly let up the slack between this knot and the descender, making sure that it holds.   To release the Quick Lock knot, merely yank on the braking end of the rope and the knot will slip loose.   This knot is normally secure if tied correctly thru an appropriate size hole, but be aware that small rope diameters and other factors can lower its reliability. **NEVER** use it to anchor a rappel rope, and don't depend on it for absolute security.

Another method of locking is to wrap the rope, in a Figure-8 fashion, around your legs (Fig. 13-8).   This isn't the most comfortable thing in the world, but it is safe.

Fig. 13-8

Securing methods can also be used for locking a rappel (see next section: "Securing a Rappel"). The methods where the securing device is above the descender usually work well for this. They can be unlocked easily by standing up in a loop formed in the braking end of the rope (Fig. 13-9).

Fig. 13-9

SECURING A RAPPEL

Securing, Safeguarding, and "Dead-Man" System are terms describing ways to provide instant braking or locking of a rappel, if the rappeller should lose control. These terms are also used for methods to prevent rappelling off the end of a rope.

Securing a rappel is wise in many cases, particularly in cave rappelling where falling rocks are a problem. However, securing techniques are normally a poor substitute for a top belay. They can make getting over edges hazardous, because of their tendency to lock at unexpected moments. They can also pose problems in tactical operations, where speed is important.

The most common way of securing rappels is to attach a Prusik knot to the rope – above the descender – and connect it to the harness carabiner via a short sling (Fig. 13-10). During the rappel, the balance hand grips the knot, keeping it from locking. In an emergency, if the balance hand loses its grip, the knot will lock and quickly stop the rappel – maybe!

Unfortunately, if you are moving fast, and the knot doesn't lock instantly, it may melt from the friction. A knot formed from thin cord can melt from sliding only one foot! Its almost impossible for knots to grip instantly, and people have been killed because of it. For maximum safety, securing knots should be made from a flexible, low stretch cord, like polyester. The cord should be at least ¼ in. (6mm) in diameter, but not so large that the knot cannot tightly grip the rope. The knot must contain enough turns so that it will not slide down the rope when heavily weighted (see Chapter 12 for more information).

Another problem with this, as with most securing methods, is the danger of "freezing" and not letting go of the knot. When this happens, you simply ride the rope to the ground – swiftly! This is perhaps the biggest danger of securing methods.

A safer way to use a securing knot is to put it on the braking end of the rope and control it with your braking hand (Fig. 13–11). Attach it to a sling tied around one thigh and to your harness. Since much less force is required to stop the rappel at the braking end of the rope, the knot should not melt.

Fig. 13-10

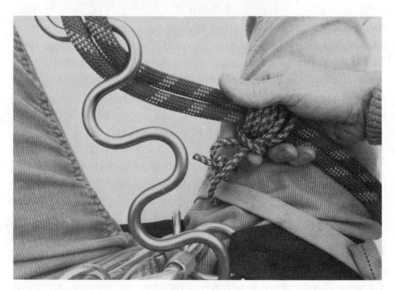

Fig. 13-11

Just make sure that the knot cannot get pulled into the descender when it locks. This could cause the securing system to fail, or could jam the descender. Another reason to put the knot on the braking side is so if you "freeze" and grip the knot so tightly that it can't lock, you will still stop the rappel.

Rappels can also be secured with mechanical ascenders, like Gibbs and Shunts (see Chapter 12). These devices grip solidly and quickly, and usually do no damage to the rope. The simplest method of securing with a Gibbs is shown in Fig. 13-12. The thumb of the balance hand holds the cam open so that the device does not lock. The instant the hand is removed, the Gibbs will lock, and stop the rappel.

To work effectively, mechanical ascenders must not be used on ropes larger in diameter than they were designed for. They must never be used to suddenly stop heavy loads, because under these conditions the rope can be cut. Like knots, mechanical ascenders may be best used on the braking side of the descender. Placed there, they work better and do less damage to the rope.

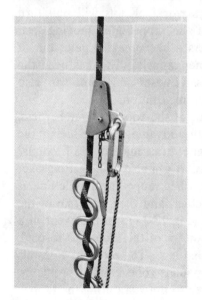

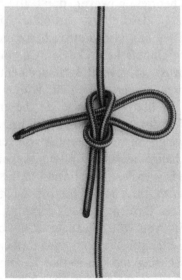

Fig. 13-12   **Fig. 13-13**

When securing devices lock, or get yanked out of reach, getting them unlocked can be difficult. Some can be released under load; others can be rigged to do so. With most, however, it is necessary to remove your weight, or to cut or untie the sling going to the device. If the securing device is attached above the descender, the weight on it can be removed by standing in a loop of the rappel rope, as shown in Fig. 13-9. Weight can also be removed by using a prusik and foot sling. If these techniques prove impractical, the sling to the device can be cut.

One way to unlock is to tie the securing device's sling with a Slipped Sheet Bend (Fig. 13-13), or other knot which can be untied under tension. Be very certain that this knot is tied correctly – and back it up with Overhand knots. To undo the Slipped Sheet Bend, simply pull on its Slipped end.

Securing a rappel also means preventing a fall from the end of the rope – in case it doesn't reach the ground. A popular method involves tying a large knot in the end of the rope. In theory, the knot will catch in the descender and stop the fall. Theories, however, don't always work! Knots will pull thru large descenders, particularly if the rappeller is moving fast. They can also destroy descenders – like breaking gates out of carabiners in Brake-bar/Carabiner rappels. This has been the cause of several fatal accidents. End-of-line knots can also break fingers! Use them with extreme caution.

A better way is to tie the end of the rope to your harness. This is very reliable and has the advantage of reducing the weight of rope below the descender, thus making less frictional change necessary on long rappels. It can also make getting over edges easier. A variation of this, when rappelling double-line, is to tie the ends of the ropes together, and then pass one of the lines thru the harness carabiner (Fig. 13-14). In the event of a fall you will be caught at the bottom of the loop thus formed. Better still, tie both ends to the harness. The problem with this securing method is that it compounds rope kinking problems, because the rope is no longer free to untwist during the rappel. Its use can be justified, however, on the grounds that it certainly beats falling!

Fig. 13-14

Even if your rope does reach bottom, this securing method can be useful. Tie into the rope high enough above the ground to prevent your hitting it, in case of a fall. The rope can be untied from the harness carabiner when you reach that point, and the rappel continued. Just remember to allow for rope stretch; lest you hit the cold, hard ground because of the "yo-yo" effect.

CONNECTING ROPES

If the rappel is longer than your rope and there is no place to reanchor on the way down, the only recourse is to connect two or more ropes together. Tying them together before descending is fine if your type descender will pass knots. Most descenders won't.

If it is necessary that two or more ropes be tied together, getting around the knot(s) can be accomplished by the following method: Rappel down to where your descender is a short distance above the knot and attach an ascender to the rope, above the descender, in a manner similar to that shown in Figs. 13-9 and 13-12. Via a strong sling, connect the ascender to your harness and carefully transfer weight to it by rappelling down a few more inches. Your descender should now be slightly above the knot. With tension no longer on the descender, it can be removed and reattached to the rope below the knot. Using another ascender and a foot sling, tension can be taken off the first one, transferred to the descender, and the rappel continued. One ascender could be used for this process, and it could be released by cutting or untying the sling, but for safety reasons this is not recommended. Be careful about stopping too near the knot, lest it get jammed in the descender while you are transferring weight to the sling – and use only new and strong sling which will not break.

Another method, useful with some descenders, is to feed the bottom rope up thru the descender, and connect it to the upper rope with an ascender (Fig. 13-15). For safety and simplicity, tie a large loop in the end of the rappel rope before throwing it over. During the transfer process, simply stand in this loop while threading the bottom rope up thru the descender. The bottom rope can be carried in a rope bag, and once connected can be tossed to the ground or payed out as you rappel. Tension transfer, from one rope to the other, is done by removing your foot from the loop and starting the rappel again. Untie the foot loop before the descender reaches it, and tie a knot in the end of the rope after it passes out of the descender (Fig. 13-16). For extra safety, put another ascender on the rope as backup.

If this method is to be used with ropes already tied together, an ascender and foot loop should be attached to the upper rope a short distance above the knot. This will take the place of a loop in the end of the rope. The lower rope can then be untied, passed up thru the descender, and attached to the ascender you are hanging from, or to another ascender. Using a separate ascender – plus an extra one for backup – is the safest way to go.

Fig. 13-15   **Fig. 13-16**

The fastest method of transfer is to have a rope faked in a leg pouch (Chapter 2, "Rope Handling"), with a descender and ascender already connected (Fig. 13-17). To use, simply attach the ascender to the rope you are rappelling on, above your descender, clip descender #2 into your harness carabiner, and continue rappelling. For maximum safety, it might be wise to put a knot in the end of rope #1, in case the ascender should start slipping. This method is best accomplished if the ascender is a type which can be easily attached to the rope with one hand.

Fig. 13-17

DEPTH FINDING

When you are going over an edge in the dark, or in rough mountain areas, it may be impossible to visually determine if the rope reaches the ground. Unless you wish to climb back up, its best to make sure – before rappelling.

The most commonly used method is to toss over a rock and count the seconds it takes for it to hit bottom. This is called the Timing Rock method, and is only fairly accurate. Difficulties in timing and hearing, air density, rock size and drag coefficient, echoes, and a number of other factors make this a method not to be trusted completely. In fact, if you can get even 10% accuracy you will be doing extremely well.

The basic formula for timing is Depth (in feet) = Seconds^2 x 16. This is theoretical, and gives a number which is greater than the actual depth. It gets more inaccurate as depths get greater, but below 200 feet it usually provides reasonable accuracy. A stop watch must be used for the timing, and the rocks must be at least four inches in diameter and fairly smooth.

RAPPELLING THRU WINDOWS

In certain operations it is necessary to bounce into window openings or holes. Depending upon the size of the window, and what is inside, this type rappel can be easy or dangerous. Danger in tactical situations can be minimized by rappelling to the side of the window and ascertaining what is inside – before going in. Because of the hazards of window rappelling, all practice should be supervised by experienced instructors.

The basic rule for window rappelling is to go in horizontally and fast, and as near as possible to the bottom of the opening (Fig. 13-18). Coming in low prevents you from flipping into a vertical position and smashing your face against the wall. All such rappels should be done with a full face shield attached to the helmet. Avoid having too much friction in the rappel system, because once you are halfway thru the window, you must be able to release friction instantly and slide on in. Window rappels must never be done thru glass – this only works on TV!

AUSTRALIAN RAPPELS

Australian (face down) rappels are used when it is necessary to have full view of the terrain below. They are seldom very pleasant, but can be of absolute importance in combat and tactical assault situations, where it is sometimes necessary to rappel and fire a weapon at the same time.

Fig. 13-18

Some harnesses can be clipped into on the backside for doing Australians, but the military harness is very handy because the carabiner can be attached on the side and slid to the back when its time to rappel (Fig. 13-19). This way, if the descender jams, you may be able to rotate your body to where you can fix things.

Fig. 13-19   Fig. 13-20

Australian rappels can be extremely painful, and hang times are very short, especially with a military style harness. Comfort and safety are greatly improved if the waist strap of the harness runs across the pelvic bones instead of the stomach.

UPSIDE-DOWN RAPPELS

There is little practical use for rappelling upside-down; however, every rappeller should try it in order to be comfortable with the feeling (Fig. 13-20). One never knows when he might accidentally be flipped upside-down. If stability is necessary during the rappel – as in certain combat situations – try wrapping a leg around the rope as you slide down. This may keep you from swinging about. Don't worry about falling out of the harness; if you are wearing one of any quality, this cannot happen. If you can't rappel upside-down safely, you can't safely rappel!

HELICOPTER RAPPELS

Sometimes it is necessary to rappel from helicopters. When doing so, remember that they are sorry anchors – when compared with a good tree. Because of air turbulence, the helicopter can accidentally pull up just as you are ready to hit the ground. Unless you brake instantly, you may rappel off the end of the rope – now far above the ground.

It is very important to use static rope when rappelling out of small helicopters, and to rappel smoothly. Rappelling on stretchy dynamic rope, and stopping quickly, will result in a series of yoyo-like bounces which can throw a small copter out of control. With big helicopters, it matters little what you use. For info on how to jump out of a helicopter, see Chapter 11.

GETTING UNJAMMED

If your descender jams during a rappel, you may have to remove your weight from it in order to move again. Jams can occur from hair or clothing caught in the descender, equipment problems, or kinks in the rope. If you are hung, and forgot to pack your ascenders, remember that emergency Prusiks can be made from boot lace, strips of clothing, and even locks of long hair. For info on how to use Prusiks, see Chapter 12, "Texas Inchworm Method"; and "Locking a Rappel" and "Securing a Rappel" in this chapter. During the unjamming process, a knife or scissors may come in handy. Just don't cut the rope!

TANDEM RAPPELS

Two or more people can rappel on the same rope at the same time. The simplest and quickest method is to attach all the rappellers to the same descender. They can then descend "banana"

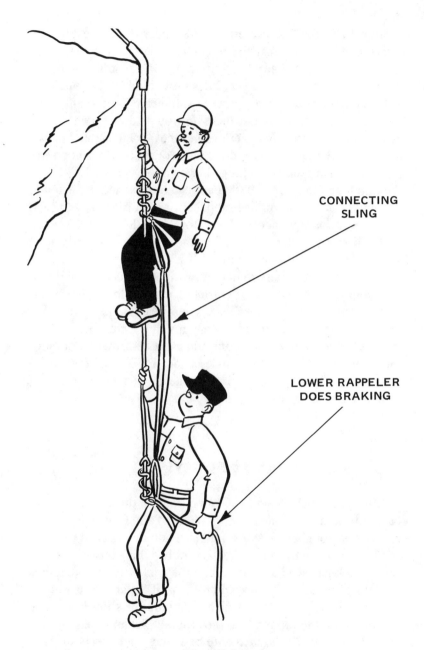

CONNECTING
SLING

LOWER RAPPELER
DOES BRAKING

Fig. 13-21

fashion, with the one nearest the descender doing the braking. This technique puts great strain on equipment, and should never be used except in emergency situations. Depending upon the descender used, it may be very difficult to hold the braking line.

If one descender cannot provide sufficient friction, and if its absolutely necessary that two loads be brought down at once, such as a rescue or equipment lowering, then the tandem method shown in Fig. 13-21 might be in order. Two (or more) descenders are placed on the rope and tied together by a long sling. This forces them to function as one. Both loads can be attached to the lower descender; or as shown, one rappeller or load is hooked to each device, and the rappeller on the bottom does the braking.

It is best if the descenders are of the same type, and of a type which causes little rope kinking. If a load is attached to each descender, it is very important that the upper descender be rigged so as to provide equal or greater friction than the lower descender, particularly when the upper load is heavier than the lower load. During the rappel the upper descender will get much hotter than the lower one, so if you are controlling the descent, do not judge the temperature of the upper descender by that of the lower. Such an error could be fatal! Whenever you rappel tandem, assume that the descenders are very hot – and go slow.

ROPE RETRIEVING

After you rappel down, you can do one of three things: 1. Go back up and unhook the rope; 2. Walk off and leave the rope; 3. Retrieve it from below. Most rappellers do the latter.

The simplest retrieval method is simply pulling on one end of a rope which has been looped around the anchor for a double-line rappel. There are few retrieval problems if the anchor is smooth and near the edge, and the rope has no kinks or knots in it. You just pull until the rope falls to the ground. Such rappels are safe and reliable if you're hooked onto both lines, and they both reach the ground, and are of the same diameter.

Fig. 13-22   Fig. 13-23

If the anchor is rough, the rope can be run thru a rope protector which is positioned around the anchor. This makes retrieval much easier. In bad situations, a strong sling can be placed around the anchor, and the rope run thru it. This method can be used to move the anchor point near the edge, which also makes pulling easier. Position the sling so that it does not run across sharp edges or bind the rope against rough rock. Always use a carabiner (Fig. 13-22) or descending ring (Fig. 13-23) on the sling if the rope is wet, since wet ropes can be difficult to retrieve. Remember that ropes loose strength when bent around a tight radius, so it might be wise to place several rings or carabiners together on the sling to reduce the sharpness of the bend. This will make retrieval easier and the system safer.

If you have only one rope, and must rappel over half its length, it can be retrieved by the use of a Reepschnur system. As shown in Fig. 13-24, the rappel rope is run thru a strong chain link which is attached to the anchor line. It is then tied to a small cord, called the Reepschnur cord.

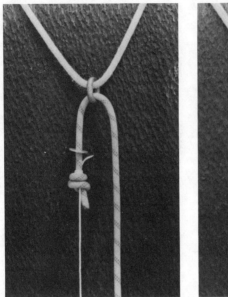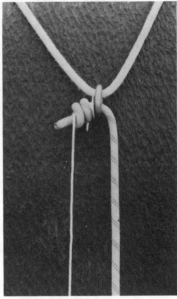

Fig. 13-24

During the rappel, the rope is prevented from pulling thru the chain link by the large knot in its end. When the rappel is done, the rope is retrieved by pulling it down with the Reepschnur cord.

If quality parts are used in the Reepschnur, this method is very safe and reliable – the main danger being that of the knot pulling thru the chain link. This can easily be prevented by placing a large washer on the rope, between the knot and the chain link (as shown in the photos). The hole in the washer should be slightly bigger than the rope, and its edges should be rounded and smooth. The chain link should be smooth too.

The best knot to use with a Reepschnur is a Double Fisherman's. This knot is large, and if tied correctly will not come loose from the small cord. To reduce the chances of the knot jamming in a crack during retrieval, it can be wrapped with adhesive tape. Before retrieval, make sure the rope has no kinks in it, and remember that pulling down a long rope requires much force. If the Reepschnur cord is not strong, it may break.

Gravity can be harnessed to retrieve a rope. The Gravity Method is rigged much like a Reepschnur, but instead of a cord, a heavy object, like a log, is tied securely to the rope. When the rappel is done, the rope is released, and the weight pulls it to the ground.

In order to prevent a cracked skull, the rappeller should carefully lower the weight as far as possible – then quickly exit the area when the rope is released. This method should be used only in places where the weight won't get hung as it falls.

Fig. 13-25

Another retrieval system utilizes a Fiffi Hook (see Chapter 7). As shown in Fig. 13-25, the rope is tied to the Fiffi, which is then hooked onto an anchor sling or descending ring. As with the Reepschnur, a small cord is also connected to the Fiffi. When the rappel is over, a slight pull on the small cord will cause the Fiffi to unhook, and the rope will fall to the ground. If the Fiffi is of good quality and hooked properly, this is a reliable method. During retrieval, however, the Fiffi could hook onto a tree or rock as the rope falls.

The retrieval cord is attached to the Fiffi via a small eye at the top. For reliable unhooking, it is important that the loop in the end of this cord extend about halfway down the Fiffi's backside. The Bowline is a good knot to use when tying this loop. Use only the strongest Fiffis! If there is any doubt about strength, tie several together so that they function as one; or better yet, use another retrieval method.

The retrieval methods just described are quite safe if done correctly. The rope will remain secure during the rappel, unless something breaks. There are retrieval methods which can come undone during the rappel. These methods are extremely dangerous! For that reason there was substantial hesitation about showing the Slippery Hitch. However, since it is used by some rappellers, and has been shown on television, it must be mentioned.

The Slippery Hitch (Fig. 13-26) and its cousin, the Precipice knot (Fig. 13-27), have been used for centuries by seamen for securing lines on board ship. In these knots, the standing part of the rope grips the end and can provide enough security for a rappel. The "knots" come loose when tension is removed and the rope lightly shaken. When used by individuals who have had years of experience with knots, the Slippery Hitch can be "reliable." Unless you have this "feel" for knots, or are willing to get it, leave this method alone! It is extremely dangerous!

The Precipice knot, which is just a Single Hitch with an Overhand knot at line's end, is a bit safer than the Slippery Hitch. Both, however, can catastrophically release if not perfectly set, allowing you to fall. If tied so they are very secure, they will not shake loose readily.

Fig. 13-26   Fig. 13-27

Even when set perfectly, a slight change in the angle of the rope, while going over, can cause the "knot" to slip. Changing the angle during the rappel, by bouncing off walls, can also cause it to slip. At best, this is never a desirable method of rope retrieval. It should be used only in emergency situations, when the rope must be retrieved, and no better system can be rigged.

Before beginning the rappel, put weight on the system to be sure that the "knot" will not slip. While descending, rappel as smoothly as possible, and always maintain tension on the rope. If tension should be removed, even for a moment, the "knot" can shift. This could prove fatal when recommencing the rappel. Generally, the rougher the surface of the anchor, the safer this system is; however, this same roughness can prevent reliable rope retrieval. Retrieval is made most reliable if the end of the rope coming from the "knot" is left quite long and hanging free. Because the Overhand knot in the Precipice knot may get caught in a crack as the rope is pulled down, it may be wisest to use the Slippery Hitch if there appears any chance of jamming.

GETTING BACK

On caving and mountaineering expeditions, and combat missions, many separate rappels may have to be made to get where you wish to go, and you may have to return the same way. This can be accomplished by leaving a rope in place at each drop, so that there will be a way back up. This works fine if there is enough personnel to carry the ropes. You could get by with fewer people by using thin ropes, like 5 to 9mm; but these thin ropes can be dangerous because of excessive stretch and low breaking strength.

You can, however, do the job with one standard rappel rope and a number of small cords. Fig. 13-28 shows a small cord tied to the end of a rappel rope, which is being pulled down from around the anchor after a double-line rappel. The small cord is the same length as the rappel rope, and is left around the anchor once the rappel rope is down. The ends of the small cord are always tied together, lest it fall or get pulled from around the anchor.

When you need to go back up, merely tie the rappel rope onto the cord and pull it back into place around the anchor. This works great if the anchor is large in diameter and smooth. It becomes tricky when trying to pull the rope thru a descending ring. The best knot for pulling purposes is a Fisherman's which has been wrapped with adhesive tape to provide a smoother shape.

This method of reattaching ropes to anchors can also be used with the Reepschnur Retrieval Method. Attach to the rappel rope a cord, twice its length, and leave it looped thru the chain link once the rope has been retrieved. The best way to attach the cord to the rappel rope is to lap the two and secure them with adhesive tape. Don't use a knot; it won't pull thru the chain link.

By carrying enough cord you can do many rappels – while carrying little weight. Just make sure the cord is strong enough, and that the connection lap is long and taped tightly. You shouldn't be able to separate the ropes by hand. Things work better if the cord is of a low stretch variety, with a smooth exterior. When pulling, pull gently, lest the lap jam or the ropes separate – and don't forget to tie the cord's ends together once you get the rappel rope down.

Fig. 13-28

MULTIPLE PITCHES

The sections on Rope Retrieval and Getting Back describe techniques for going down a long drop in several stages, i.e., multiple pitches. These pitches are necessary if your rope is not long enough to do the entire drop in one rappel. In most situations, the pitches are truly separate, and the rappeller can walk about between each one and look for new anchors. However, in some situations, the drop is straight and the anchors for the different pitches must be set up while on rappel. These are situations where locking a rappel is an absolute necessity.

After you are locked in, your hands are free to establish a new anchor by placing chocks, pitons, or whatever. Sometimes one object, like a rock outcrop, will prove strong enough to put a sling around. Just make sure that the sling can't slip off. Generally, its best to use two or more anchors and tie them together in a balanced system (see Chapter 9). Since most multiple pitch accidents are caused by anchors pulling out, always test the anchor while you are still on rappel. Do this by attaching a sling to it, putting your full weight in the sling, and jumping up and down. If the anchor breaks, you will be left hanging on the rappel rope – instead of dead.

Once the anchor is established, you can hook your harness carabiner to the previously attached sling, and hang from it while pulling the rappel rope down for the next pitch. The rappel rope is then attached to this new anchor. Once it is in place, attach your descender and prepare to rappel.

There are a number of ways to disconnect yourself from the anchor, so that the rappel can commence. The safest way is to stand in a etrier while unclipping from the anchor. Whatever means is used, weight transfer from the sling to the rope should be done gently, so as not to put excessive load on the anchor.

Multiple pitches on cliff walls can be extremely dangerous. Do not attempt it until you are expert at placing chocks, pitons, or etc. Such devices are rarely "bombproof" and do tend to break loose easily. Likewise, do not judge rock outcrops or horns by their size. Pieces bigger than you may break easily.

PSYCHOLOGY

Although designed for use by all rappellers, this book is oriented primarily toward the needs of rappel instructors. Most everything of practical importance about rappelling is to be found here. However, just as important as technical knowledge is knowledge of the psychology of rappelling.

In sport rappelling, where the object is fun, there is seldom any consideration of the interaction of the mind with equipment and situation. Sport rappelling is generally done in good weather, and the mood is one of thrill and joy. It is this lighthearted mood which can sometimes lull rappellers into false security – and perhaps tragedy.

In other worlds of rappelling, the mood is seldom so joyous. Mountaineering, rescue, and combat rappels are not done for fun, and one's mind is not always clear. In these situations anxieties and phobias are magnified, and small mistakes can quickly become big mistakes. Instructors should be aware of how thought processes are affected by training techniques and equipment condition, so that anxieties produced during "real world" situations can be minimized.

Many rappel instruction programs require students to descend an extreme height on their first rappel, sometimes physically forcing them over the edge. The purpose of this is to "train out" anxieties, teach discipline, or quickly achieve the goals of the training program. There are perhaps occasions where such methods are justified, but normally they should never be used.

Although anxieties and phobias can sometimes be trained out by "shock" experiences, they can sometimes be made worse, trained in, or trigger other adverse psychological reactions. Few people with fear of snakes can be "cured" by throwing snakes on them! "Shock" techniques can also lead to physical injury, even when carried out under the most careful conditions.

"Trained in" anxieties are not always initially evident. They can surface later when an individual faces an extreme stress situation: "The night is cold and dark; you and your partner have been in the mountains for twelve hours looking for a lost child. For the last two hours you have been fighting your way thru thick brush and over slippery rocks. You are so tired you can hardly put one foot in front of the other; your partner is just as fatigued. As you come to the edge of a 200 foot cliff, you see something lying in the snow below. Could it be the child? Its hard to think straight and your fingers are almost numb, but you rig for a rappel. Your partner clips in and starts over the edge, but he is distracted for a moment by the memory of his first rappel. It was very high and there were moments of terror. His fatigue and anxiety magnify this image of terror. During this moment of distraction his attention is taken from where he is placing his foot. He slips and falls sideways only a few inches, and his rope is cut. And he falls."

Perhaps if this hypothetical person had not had a bad experience on his first rappel this hypothetical tragedy would not have happened. Stress in such situations can result from physical fatigue, low blood sugar, allergies, injuries, and a number of other factors. If anxiety and fear – from past experiences – are added to this stress load, the mind can malfunction when it is needed most. Fear causes the brain to release noradrenalin (norepinephrine), which in turn causes blood to be diverted from the cerebral cortex to the more primitive "emotional" section of the brain. This reduces logical thinking capacity and increases the chance of a rappeller making fatal errors.

To avoid "training in" anxieties, a rappeller's first descent should be no higher than that with which he is comfortable. If he feels best rappelling from a two foot cliff – start him on one. If possible, have him "rappel" down a gentle incline, and then work up from there. This method takes longer, but there is no safer way to teach rappelling. Equipment and technique errors are rare. If there is anything bad to be said about the slow, safe method, it is that it does little to teach an individual how to cope with unexpected emergency challenges.

Equipment can also be a source of fear. A rope of normal diameter can look dangerously small to an individual on his first rappel. Even the color of a rope can be a factor. If possible, phobic beginners should be allowed to use whichever ropes, harnesses, and etc. make them feel the safest. Education into practicality can come after they get over their initial fear.

Besides using sturdy equipment and gradual changes in height to eliminate fear, there are a number of subtle things which might be used to distract the beginner's attention from his anxiety. Once safely over the edge, the individual can be directed to stop the rappel and touch his nose with his balance hand. This momentary distraction may block the anxiety long enough so that he can rationally examine the situation. However, removing the balance hand from the rope should only be tried if the rappeller is hanging in a properly balanced harness and is being belayed. Other distractions which might prove useful are: tapping the nose ten times, touching the nose and then the ear, counting backwards, reciting one's address, etc.

An excellent method for "training out" fear is to let the beginner belay an experienced rappeller as he descends – using a standard descender for the belay device. Once the individual has learned to control the belay system, the rappeller should allow himself to be lowered to the ground by the beginner. This is perfectly safe, because if the person on belay makes an error, the rappel can be continued in a normal fashion. After a few minutes as belayer, the beginner will perhaps gain confidence in the strength of the system and be willing to try rappelling, using the same descender with which he did the belaying.

Visualization is another way to conquer fear, and can be used in the safety of one's home, before any rappelling is done. In quiet and comfortable surroundings, with the eyes closed, the beginner should visualize a positive rappel scenario. Like watching a "movie" in the mind, he should imagine himself safely getting over the edge, and slowly sliding to the ground. The "movie" should contain as much detail as possible, and always end with feelings of triumph. This visualization should be replayed as many times as necessary to effect a positive mental change.

Many individuals always retain fear no matter how slowly and carefully a training program is conducted. Some are forced to rappel because of job requirements, and have resentments on top of their anxieties. If this poses a danger to team performance, and if the individual must be involved in rappel situations, other means of controlling the fear can be considered.

With the aid of a physician, very deep phobias can often be trained out with chemicals, like beta blockers, which can reduce the physical effects of fear. However, all such drugs can have serious side effects, so this technique should be considered only when all other methods have been exhausted – and only if the individual has no anxiety about taking the medication.

Neither beta blockers, alcohol, or any consciousness-altering chemical should be in the body of an individual in a real rappel or rescue situation – where life depends on clear thinking. Over 95% of rappel accidents, in certain areas, involve people with alcohol or drugs in their blood.

Survivors of great tragedy almost always testify that control of fear is, for the most part, a spiritual thing. Fear of rappelling is no different. Rappellers with faith in God have overcome their fear by simply learning to trust God. There is no point in claiming faith if it is not put to practice. Dr. Norman Vincent Peale writes, "Faith is stronger than fear. So substitute faith for fear and see what happens."

Using a top and a bottom belay when training beginners is as much a part of eliminating fear as it is good safety. Just be sure the belays are set up and managed correctly. A belayer with a rope in one hand – and a beer in the other – accomplishes nothing toward alleviating a beginner's anxiety. As skill in rappelling is gained, the belay should be eliminated at times in order to build self-confidence. After all, on that cold, dark night on a mountain rescue, there may not be a belay available. If the rappeller is not comfortable without a belay, this fear can lead to an accident. Rappelling without belay is not a good practice, but it is absolutely necessary to the development of competent rappellers and rescue squads.

Equipment condition figures greatly in rappel psychology. A frayed area on a rope, or a grease spot, may go unnoticed during practice, where there are good belays and happy faces. But on that cold, dark night, when your energy is almost gone, the sight of an imperfection in the equipment may be all that is needed to increase your anxiety level to the danger point. The last thing anyone needs on a rescue is the worry of equipment failure.

In order to prevent such anxieties, it is important that all equipment used for rescue be the best available, and that it be maintained in optimum condition. All equipment should be stored in an appropriate – locked – area anytime it is not being used by the group. No individual, under any circumstances, should be allowed personal use of the equipment. The individual in charge of the equipment should be the most responsible member of the group, and a log should be kept of equipment use. Everything should be inspected on a regular basis.

Individual items like harnesses can be made or fitted for each team member, and each individual can be in charge of his own equipment. All such personal equipment should be inspected on a regular basis by the group leader. Using only the best equipment and inspecting it regularly will eliminate many anxieties and make all rescue operations safer.

In many emergency situations it will be necessary for untrained individuals to rappel. This is seldom an easy task. The fears of non-rappellers can be lessened to some extent if they are belayed from both top and bottom, but this is not always possible. It is these situations where large diameter ropes and extra sturdy equipment prove of value. Its much easier to get a non-rappeller to descend a 16mm rope than one of 11mm.

If an individual absolutely refuses to rappel alone, a tandem rappel might be in order. Children, in particular, are much more comfortable when attached to an adult. Every rescue squad should have periodic training sessions on tandem rappelling, using people with wide differences in body weight.

Concerning training sessions: There is a very wide variation of attitudes on the part of those being trained to rappel. Some, as discussed earlier, come with many fears and inhibitions, and it is necessary to proceed with them very slowly. "Though the spirit is willing, the flesh is weak," thus some will never be able to complete the program no matter how hard they try. Some good talent is lost this way.

Other individuals, particularly young males, enter training with much more enthusiasm than is good. They are usually at the head of the line, with visions of Tarzan whirling thru their heads.   Those with high intelligence usually listen well, take orders, and control themselves in a responsible fashion.   Others listen only long enough to understand basic methods, and then start to fly on their own.   This latter type can prove extremely hazardous to themselves and other members of the group, so should be dealt with immediately.   If they continue with a bad attitude, and seem disinterested in learning the details of safe rappelling, they must be asked to leave.

In general, rescue squads should be made up of cautious, conservative, mature, intelligent people who are there because they wish to be of service – not because of an ego trip.   Its usually better to have people with a little fear and lots of dedication, than egotists who have no fear and no experience in managing fear.   When emergencies arise, those experienced in managing their emotions usually fare better than those who have never had to try.

To instructors: You've taught a lot of people, and you're good at what you do. That's great, except for one thing. Its easier now than it used to be to believe that you know it all. When someone offers a suggestion, you feel something inside you rebel at the thought of "being told." After all, you're the expert.

Such feelings are normal, but are something which must be continuously fought. Its so easy to fall into the "know it all" rut, but there is no other attitude which can so quickly ruin a good instructor. Always go into each new program expecting to learn things from your students. Each student should be encouraged to think, create, and add to the total learning experience. More often than not, new concepts will emerge from such a comfortable and intellectual atmosphere.

"Why learn to rappel?" Because few groups will ever need to rappel in order to effect a rescue, this question is often posed. The question is also valid for non-rescue groups, like Scout troops. There are many reasons for learning rappelling: Its an excellent way to build cohesion and comradeship amongst group members, and to increase their self-confidence.   When done correctly rappelling is a very safe activity which most everyone will enjoy, and it has a way of dissolving emotional barriers.   Few things build trust more readily than one individual being belayed by another. It is also a survival skill, and the knowledge of how to rappel might perhaps save a life in a bad situation.

There are many reasons to learn to rappel. There are few, if any, reasonable reasons not to learn. So why rappel? Why not!

SAFE RAPPELLING

GLOSSARY

Aardvark – A burrowing African mammal that feeds on ants and termites. Not known to be skilled at rappelling.

Abseiling – Rappelling.

Anchor – The support (tree, piton, etc.) to which the rappel rope is attached (anchored) at the top of a drop (wall, cliff, etc.).

Ascender – Any instrument used for ascending ropes which "clamps" on to the rope, providing movable hand or footholds.

Balance Hand – The hand which is placed on the rope, above the descender, solely to maintain balance – not support weight.

Belay – A rope-oriented safety system used to prevent a rappeller from falling if he loses control or if his rope should break.

Belayer – One who controls or operates a belay system.

Bend – A knot which unites two rope ends. Something a steamboat comes around.

Body Rappel – A means of rappelling wherein friction for speed control is obtained by wrapping the rope around parts of the body.

Bombproof – The term applied to an anchor which will remain secure under any extreme to which it might be subjected.

Braided Rope – A type of rope which is constructed by braiding or weaving the yarns, instead of twisting them.

Braking End – The portion of the rappel rope below the descender; also referred to as the "tail."

Braking Hand – The hand used to control the descent speed of a rappel; normally gripping the rope below the descender.

Carabiner – A device used to quickly couple two or more ropes or pieces of equipment.

Chest Harness – An upper body harness, which is normally used to balance a climber in an upright position while hanging on a rope.

Chock – A block-like device which is wedged in cracks in order to secure ropes and etc. to a rock face.

Control Hand – Braking hand.

Cord – A small diameter rope which is used for connecting things, tying Prusik knots, and etc.

Dead Man Brake – A securing device which automatically stops the rappel if the rappeller should lose control.

Descender – A device which enables a rappeller to safely descend a rope without having to support his entire weight with his hands.

Descending Ring – A small ring which is used for various rappel purposes; especially for coupling ropes in anchor systems.

Drop – The wall, cliff, or etc. to be descended. A small quantity of liquid that is somewhat spherical or pear-shaped when falling.

Dynamic – Anything which is moving or active; as in the forces generated by a rappeller descending a rope, i.e. kinetic energy.

Ears – Sound-reflecting appendages protruding from a rappeller's head. Small rope-guiding horns on a descender.

Etrier – A short rope ladder used in rock climbing.

Faking – Coiling a rope or cable. In rappelling, it denotes coiling a rope in a zig-zag fashion; usually inside a bag.

Falling – The verbal signal, from rappeller to belayer, that he is falling.

Fiber – Threadlike structures which are twisted together to make yarn and rope.

Grapnel – An instrument with hooks or claws for grasping; it can be used for casting and securing ropes to high objects.

Harness – Any strap-like device which fits on the body, enabling the wearer to be suspended from ropes, anchors, and etc.

Harness Carabiner – The carabiner which is used to connect a descender to the rappeller's harness.

Hitch – A knot which makes a rope fast to another rope or object, and which can be readily untied.

Kernmantle Rope – A composite rope consisting of a braided outer sheath covering a fiber core.

Kinetic Energy – The energy of a body which results from its motion; equal to mass times velocity squared, divided by two.

Laid Rope – A type of rope constructed by twisting two or more yarns together, in a spiraling fashion.

Locking – A means of fixing one's position on a rope during a rappel, so that the hands are freed for other tasks.

On Belay – The verbal signal, from belayer to rappeller, that it is safe to begin rappelling.

On Rappel – The verbal signal, from rappeller to belayer, that he is starting to rappel.

Off Rappel – The verbal signal, from rappeller to belayer, that he has safely reached the bottom of the drop.

Piton – A spike-like metal device which is driven into cracks in rock, in order to secure ropes and etc.

Platypus – An Australian egg-laying mammal with webbed feet and poisonous spurs; not generally known for its rappelling ability.

Rappeller – One who rappels. Commonly referred to as "crazy" by non-rappellers.

Rappelling – A controlled means of descending a rope, in which one's hands must support only a fraction of his total weight.

Rock – The verbal signal, from those above to those below, that something dangerous is heading their way.

Rocks – The plural of rock!

Rope – The verbal signal given before a rappel rope is cast over the edge. Also, a long, flexible instrument for rappelling on.

Seat Harness – A harness which fits around the buttocks and thighs, enabling a rappeller to hang in a sitting position.

Securing – Methods utilized by a rappeller to prevent himself from falling to the ground in case he loses control.

Slack – The verbal signal, from rappeller to belayer, to ease tension on the belay rope.

Sling – A loop of rope, webbing, or etc., for making anchors, prusiks, and etc. Primitive weapon used for throwing rocks.

Splat – Sound produced at the end of an excessively fast rappel.

Standing Line – That portion of a rope which can be rappelled on.

Static – Not moving; as in a weight hanging from a rope.

Tail – A rappeller's posterior or dorsal area. Also, that end of a rope nearest to the ground.

Tension – The stress produced by a pull of forces on a rope. Also the verbal signal to "take up slack" in the belay rope.

Thimble – A grooved metal ring inserted in a loop of rope to prevent wear.

U.I.A.A. – International Union of Alpine Associations, which sets safety standards for equipment used in mountaineering.

Ultraviolet – The invisible, shortest wavelengths of light, which cause sunburn and deterioration of plastics and cloth.

Yarn – A thin cord-like structure made by twisting together two or more fibers; used for constructing rope.

Zy-glow – A test technique which uses ultraviolet light to reveal cracks in metal, after saturating the metal with fluorescent oil.

Zygodactyl – Having the toes arranged in two opposed pairs, two in front and two in back; as in the feet of a parrot.

SOURCES

Information on rope and rappel equipment standards can be obtained from the following sources:

National Fire Protection Association
Batterymarch Park
Quincy, MA 02269

Union Internationale des Associations d'Alpinisme (U.I.A.A.)
c.p. 237
CH 1211 Geneva 11, Switzerland

National Speleological Society
Cave Avenue
Huntsville, AL 35810

British Mountaineering Council
Technical Committee
Crawford House, Precinct Centre
Manchester M13 9RZ England

International Association of Fire Fighters
1750 New York Avenue, N.W.
Washington, D.C. 20006

Cordage Institute
1625 Massachusetts Ave., N.W.
Washington, D.C. 20036

Information on equipment shown in this book can be obtained by contacting the manufacturers listed in this section. Some of these companies also sell equipment and provide rappel instruction.

American Rescue Systems
P.O. Box 1776, Zephyr Cove, NV 89448

Arova Lenzburg AG (Mammut)
CH-5600 Lenzburg, Switzerland

Bry-Dan Corporation
P.O. Box 295, Moraga, CA 94556

California Mountain Company (CMC)
P.O. Box 6602, Santa Barbara, CA 93160

Camp, s.p.a.
Via Roma 23, 22050 Premana (CO), Italy

Chouinard Equipment
P.O. Box 90, Ventura, CA 93002

Colorado Mountain Industries (CMI)
P.O. Box 535, Franklin, WV 26807

Descent Control
P.O. Box 6405, Ft. Smith, AR 72906

Edelmann & Ridder (Edelrid)
D 7972 Isny/Allgau, W. Germany

Forrest Mountaineering
1136 Speer Blvd., Denver, CO 80202

Kong-Bonaiti
Cas. Post. 14, 24030 Monte Marenzo, Italy

Lirakis, Inc.
30 Greenough Place, Newport, RI 02840

Lowe Alpine Systems
P.O. Box 189, Lafayette, CO 80026

Mar-Mex International
P.O. Box 723126, Atlanta, GA 30339

Miller Equipment
P.O. Box 271, Franklin, PA 16323

New England Ropes, Inc.
Popes Island, New Bedford, MA 02740

Penticton Engineering
100 Industrial Ave., Penticton, B.C. V2A 3H8, Canada

Petzl
Z.I. Crolles, 38190 Brignoud, France

Pigeon Mountain Industries (PMI)
P.O. Box 803, Lafayette, GA 30728

Rappel Rescue Systems, Inc.
214 Little Falls Road, Fairfield, NJ 07006

Raven Products
P.O. Box 8663, Portland, OR 97207

Recreational Equipment, Inc. (REI)
P.O. Box C-88125, Seattle, WA 98188

Rescue Systems, Inc.
Rt. 2, Box RSI, Little Hocking, OH 45742

Ruapehu Mountain Equipment Co.
470 Parnell Road, Auckland, New Zealand

Seattle Manufacturing Co. (SMC)
12880 Northrup Way, Bellevue, WA 98005

Stubai
P.O. Box 31, A-6166 Fulpmes, Austria

J.E. Weinel, Inc.
P.O. Box 213, Valencia, PA 16059

INDEX

Rappelling can be a hazardous activity. Do not attempt to learn to rappel by using this book as your only source of information. This book is not a substitute for a competent instructor.